The
Left Stuff

The
Left Stuff

*How the Left-Handed
Have Survived and
Thrived in a Right-
Handed World*

Melissa Roth

M. Evans and Company, Inc
New York

M. Evans and Company, Inc.
216 East 49th Street
New York, NY 10017

Library of Congress Cataloging-in-Publication Data

Roth, Melissa.
The left stuff : how the left-handed have survived and thrived /
Melissa Roth.
p. cm.
ISBN 1-59077-081-1
1. Left- and right-handedness. 2. Human evolution. I. Title.
QP385.R65 2005
152.3'35--dc22
2005007080

Typeset and Designed by Chrissy Kwasnik

Printed in the United States of America

9 8 7 6 5 4 3 2 1

For my lefty role models,
Mary Roth and Joanne Barry

Contents

Acknowledgments

This book has traveled a very long road and owes many debts, not the least of which are on my Visa card.

First and foremost, I want to thank the scholars who've dedicated years of research to this topic and helped to advance the science. Several of them were especially generous with their time and knowledge, in particular Michael Peters at the University of Guelph, whose insights and comments and translations of complex research were invaluable. I'm also grateful to Steve Christman, Dan Geschwind, Lauren Harris, Peter MacNeilage, Clare Porac, and Jeannette Ward for taking the time to share research findings and expertise. And though we never spoke, I'm grateful for the many contributions of Marian Annett, the foremother of much of the research covered in this book.

Special thanks go to Ann Treistman, who was there at the beginning of this project and ultimately helped to give it a second life several years later. And to Elizabeth Sheinkman at the Elaine Markson Agency, also there at the beginning, for not giving up.

I'm grateful to M. Evans and Company, in particular Matt Harper, for his thoughtful and thorough edits, big and small, and for shepherding the manuscript through its grueling last laps; and to Enrica Gadler, who acquired the book and provided it with a better road map. Also at M. Evans, I'd like to thank Chrissy Kwasnik and Dina Jordan.

I was also fortunate to have as my "frontline" editor Mary Beth Ross, whose scholarly insights and sharp editorial instincts made their mark throughout this book's long life. Lucky for me she is also my mom, so I could afford her. It was her early interest in language and the brain that sparked my own curiosity about the topic many years ago.

Thanks also go to Ethan Nosowsky and Heather King for their edits and comments on the cultural history chapter and to Rachel Warren for her input on the science chapters.

For the book's pre-9/11 incarnation, I am indebted to David Hirshey, who also gets credit for the title. Jeff Kellogg provided helpful editorial guidance on early drafts. Geoff Shandler's big picture suggestions were also incorporated. I was also lucky to find Jon Marelli, a resourceful researcher and scholar.

So many people shared tips and thoughts and research findings over the years, I regret that I won't be able to list them all here. But special thanks go to Rich Roth, my father, and Geoff Roth, my brother. Also thanks to Lila Hicks and Kiku Lani for their documentary insights, Mary Christ for her hawk eye for all things lefty, and Tina Jordan for her ideas and suggestions, no doubt the product of her left-handed brain.

Finally, special thanks to my late aunt Katherine Reckling Scott, a switched lefty who bought me a special left-handed book from this very publisher when I was seven years old.

Author's Note

It was 1979, and I was in the middle of my adoption phase. My mom had just taken me to see *Annie*, and I was fairly sure she was getting ready to tell me my own past life story, the one about the orphanage and how she rescued me from all those hard knocks and loud kids. She must have thought it was a little odd the way I'd started courting all her friends, dragging them up to my room to show them my cartwheels and Cray-pas drawings. I was convinced these displays would make each of my "real moms" all the sorrier for leaving me as they did.

Psychologists will tell you that many kids go through an adoption phase, imagining a different life somewhere, where parents don't fight and dogs can sleep on your bed. My adoption phase was a little more scientific: I was sitting at Thanksgiving dinner, counting the number of forks I would need to round up for dessert duty, when I decided to count them again. Twenty-two people were gathered around my grandmother's table, and only one was using her left hand: me. My fifth-grade teacher, Mr. Burns, had just explained that there was no such thing as a "right" or "wrong" hand. Some of us are just born to be left-handed. It's passed down from our parents. My parents, and my grandparents, were all right-handed.

My kindergarten teacher, Mrs. North, regarded a left-hand preference as an act of defiance that needed to be scolded and shamed, along with biting and sand-eating. She had big, silver hair and an abrupt voice, but my only clear memory of her is the day she yelled my name across the sea of nap blankets. *"Muh-lisssa! What are you doing?!"* I was quietly coloring the Winter Wonderland mural with a crayon, but I was too scared to answer. "Why are you holding the crayon in *that* hand?"

Mrs. North was about the same age as my grandmother. When she was in school, the nuns tied my grandmother's left hand behind her back to break her of her "sinister" habit. She struggled with the early grades, all that emphasis on legible handwriting. Later on, there were some battles with the sewing machine and the iron, causing her sisters to worry about her marriage prospects. But she managed to become a strong student, a wife, and a brilliant general, brigadeering an army of seven kids and eighteen grandkids with the tactical skills of Napoleon (also a lefty). Of course, grandchildren struggle to decipher the right-handed notes she scrawls in our birthday cards. And in the fall of 2000, she contributed to the presidential election melee when she accidentally voted for Pat Buchanan. (She was an eighty-five-year-old switched lefty in a right-handed voting booth with a dyslexic butterfly ballot.) But when my grandmother first shared her own lefty tales with me, I stopped searching for my real parents. They were there the whole time, just being recessive. Or so I thought.

Vindication finally came ten years ago when I visited my first neurologist. The first question he asked, before I could tell him about my migraine headaches, was "Are you left-handed?" At *last*! I had found what I'd always suspected and secretly wished for: one all-encompassing alibi that explained everything, every headache and foul mood, all the bad luck and false starts. I was *left-handed*. "We don't know why," the doctor said, "but the left-handed are more likely to suffer from migraine headaches."

Six months later, when I discovered that simple changes in diet, sleep, water intake, and stress levels could ease my headaches, the same changes that seem to help right-handed migraineurs, I stopped accepting my doomed lefty fate and decided to look into this great mystery. After some research, I figured out what the doctor was talking about. It is described as the theory of the "left-hander syndrome." Captured in a controversial 1991 book by the same title, the theory claims that left-handedness is a "pathology," caused by prenatal conditions or birth stress that leads to a form of brain damage. It was a little bit more alibi than I was looking for, but it began the research that lead to this book.

Since then, I've spent several years staring at people's hands. I've also spent hours at the zoo, waiting for the apes to rummage for food. I've had conversations with yogis ("the left side of the body is female and lunar"), taxi drivers ("in my country, we sit on your hand,") and leg waxers ("you people, hairier left leg"). Recently, I was caught staring at a stranger's right temple, trying to imagine if his sentences were being formulated from there.

Along the way, I volunteered for brain-imaging experiments, genetic testing, hand analysis, career aptitude testing, and biomechanics studies. I've even asked mature friends to beg for autographs from suspected closet cases. I managed to stop short of digging up medieval graves to measure the lengths of right and left arm bones. Luckily, a group of British scientists was not so restrained.

I have, however, excavated decades worth of scientific studies, journal articles, comparative brain scan analyses, psychological tests, and left-handed lore. I've also interviewed neuroscientists, historians, geneticists, anthropologists, cognitive psychologists, primatologists, kinesiologists, and lefties of every stripe—they come in many.

My research has turned up hundreds of pages of valuable explanations and excuses. When I lose at Scrabble, it's because I'm left-handed (we *like* our words scrabbled). When I knocked a friend's $190 sunglasses into the ocean, it was my clumsy left hand's fault.

And when I had trouble tackling parts of this book, well . . . at least I didn't write it backward like da Vinci did.

While I am happy to share all the apologies and alibis in this book with other sinistrals—as the scientific literature likes to call us—the disordered status we might occasionally crave is not where my study *left* me. In fact, for all the bad news, insults, and obstacles that make up a lefty's life, being born left-handed may just be one of evolution's uncelebrated gifts.

1

The Left-Handed Renaissance

They pop up in studies of the dyslexic, the distressed, and the generally disordered. They are rumored to be at greater risk of immune problems and alcoholism. And according to one widely publicized study, they die several years before everyone else. So why are the left-handed suddenly doing so well?

Just a century ago, fewer than five percent of Americans were left-handed. Today, their rates have nearly tripled. Somewhere between eleven and fifteen percent of the United States population qualifies as left-handed, according to researchers. Recently, a British scholar suggested that a gene that leads to left-handedness has been surging among Westerners over the last few generations, a statement that led the *Sunday Times* of London to declare a left-handed "renaissance."[1]

Few biases have survived as much history and crossed as many boundaries as those against the left hand. From the Bible to the Koran, from primitive tribes to advanced civilizations, from the dawn of recorded history to twenty-first-century American parents, fear and suspicion of the *sinister side* is so deeply ingrained few even notice its existence. It has been held up as a symbol of awkwardness, disobedience, even evil, and for centuries, few who preferred to use their left hand dared admit to it. Stories are plentiful of "caggie handers" trying to eat or to write, only to have their left hand whipped, scalded, sat upon, shouted at, and even balled into a stump with duct tape. A skilled left hand was even used as evidence of witchcraft, and sinistrals were sent to the village bonfire.[2]

Yet throughout recorded history, left-handedness has also been a subject of curiosity. It's a matter of human interest because it's a peculiar distinction found in a small minority. It is a matter of scientific interest because it's believed to provide important clues about the organization of the brain. The hands might look and feel identical, but they develop different talents, and this is a reflection of the differences between the two hemispheres of the brain.

Brains are a fundamental part of what makes us human, so exploring their organization also raises questions about what defines our humanity. The development of a language center in the brain, for example, has come to be regarded as a defining human characteristic, one that sets us apart from the other animal species. But so has right-handedness. Homo sapiens are believed to be unique in their specieswide tendency to favor the same side. And since the great majority of humans use their right hand for fine motor skills, right-handedness has come to be viewed, in an oversimplified manner, as a characteristic unique to man.[3]

So where does that leave the left-handed?

Lore and the Lethal Lefty

"Do We Want a Left-Handed President?" the headline of a December 1988 *Washington Post* article inquired, just before the left-handed George Bush Sr. was to raise his right hand on Inauguration Day. It was the rise of the "weakest link" theory of left-handedness. Newspapers abounded with stories connecting left-handedness to accidents, illnesses, psychological disorders, and learning disabilities, ultimately leading to a shorter life span. The kickoff was a 1980 study of Canadians, led by psychologist Stanley Coren, which found a significant drop-off of lefties in older populations. A 1988 follow-up study of baseball records, also conducted by Coren, came to the same conclusion: southpaws disappear with age. "This left us with the macabre possibility," Coren told the *New York Times* in 1989, "that there is a smaller and smaller number of left-handers in the statistics because . . . they had died." [4]

Two years later, Coren published a book popularizing a theory that left-handedness was a sign of an underlying "syndrome," a red flag for a kind of early brain damage that predisposes a person to a palette of problems that could surface throughout his or her life. And sure enough, that very spring, when President Bush entered the hospital for an overactive thyroid gland, often caused by stress, one prominent medical expert instead offered this explanation for his condition: "People with left-handedness are more prone to auto-immune thyroid disorders." [5]

In the midst of all of this, lefties fought back using the best weapon at their disposal: name-dropping. This so-called syndrome, they argued, brought the world Alexander the Great, Charlemagne, Julius Caesar, Leonardo da Vinci, Napoleon, Michelangelo, Ben

Franklin, J.S. Bach, and Babe Ruth. Just before they could add centenarians George Burns and the Queen Mother, they were being dragged back into the public square, cowering beneath more ominous front-page headlines: "Left-Handers Die Younger" came out of Washington. "Leading with Your Left May Shorten Your Life" greeted Londoners at the breakfast table. "Is the World Really Safe for Lefties?" raised hackles in Los Angeles. [6]

Some lefties wrote to newspapers, asking how this new study might affect actuarial tables, and whether switching hands midlife could tack on a few years. Others snidely suggested a campaign against left-handedness, lumped in with other self-destructive behaviors such as smoking and drinking and driving. [7]

Shortened life span aside, the quandary persists for parents and teachers: can and should such leanings be squelched? Though the pressures to convert to right-handedness are no longer as systematic or coercive as they were before the 1950s, there are residual and newly inspired antipathies toward *that hand*. Psychologists have reported significant pressures against left-handedness in the current generation. In a 1998 survey, twenty-four percent of younger-generation lefties reported some attempts to switch their handedness, an intervention that can interfere with early development while leaving an anxious trace on temperament well into adulthood. [8]

Meanwhile, the shorter life span claim is still tossed around nearly two decades after it first surfaced. On this issue, the pro-left can be their own worst enemies, as was the case in the summer of 2000 when advocates on both sides of the Atlantic argued for special protection for the left-handed, claiming the group is at greater risk of premature death, and dredging up Coren's ominous findings. One of the United Kingdom's largest unions and a New York City councilman both earned international headlines, spurring a debate that spilled out into the op-ed pages of the London *Independent*. [9]

"Forgive me. I'd always thought that being left-handed was a minor personal foible, on a par with having dodgy eyesight . . . ," scoffed the

Independent's Adam Leigh. "To compare the minimal inconveniences of living in a right-hander's world with the practical difficulties experienced in everyday life by those with physical disabilities, or the sort of discrimination meted out to racial minorities, is to take political correctness to the most absurd lengths imaginable." [10]

"If a teacher shouted abuse at a child, and rapped her over the knuckles because she was dark-haired or short-sighted, there'd be a national outcry," shot back the paper's personal finance editor, Isabel Berwick. "But it was okay for my teachers to call me 'caggie handed' and hit me with a ruler when I smudged my schoolwork." [11]

Lost in the big debate was this: the studies that claimed lefties die prematurely turned out to be spurious, refuted by the National Institute on Aging, the United States Statistical Assessment Service, the American Academy of Actuaries, and investigators across the continents. [12]

Nevertheless, the "dying lefties" myth persists, adding another burden to those who have been labeled Satanic, Communist, dirty, and disobedient because of their hand preference. From their teachers to their parents to their employers, there is no shortage of modern-day slights and insults directed toward those "with their arms on backwards," as one teacher referred to her young student in the 1970s. Even within the exacting proofs of the scientific community, right-handedness is still considered a marker of a "normal" brain lateralization, and left-handedness is still referred to as a "risk." Adding insults to alleged injuries, lefties are excluded from many research studies; their variability renders them unpredictable, a screwball variable in the data set. [13]

While modern civilization has eliminated many of the more ominous threats to the left-handed,[14] the material world still conspires against them. The average lefties may not even notice, but in any given day, they may do battle with a hair dryer cord, get into a scuffle with a right-handed zipper, struggle while opening the lid of a jar, break a few pencils in the sharpener or smudge the ink of the pen resorted to as a result, tangle themselves up in a right-handed office phone cord while pushing buttons with their right hand, carve a notebook spiral into their left wrist as they take notes, bump elbows with their boss over lunch, calculate an Excel spreadsheet using their right-handed keyboard, cut and paste with a right-handed mouse, videotape their daughter's soccer game with a righty camcorder. After driving with a right-handed stick shift down a right-sided street to their right-handed house, they might wrestle with the kitchen implements (ladles, peelers, can openers, corkscrews, serrated knives), appliances (microwave, coffee maker, refrigerator), and tools (screws and screwdrivers).

Despite the obstacle course laid out in their path, the left-handed appear to be having a heyday. While their numbers are gaining in the population at large, they also show up disproportionately in lists of newsmakers and record breakers. They include the richest woman in the world (Oprah Winfrey) and the richest man (Bill Gates), some of the most attention-getting leaders (Bill Clinton, John McCain, George Bush Sr., NYC Mayor Mike Bloomberg, George Pataki), film and TV stars (Julia Roberts, Nicole Kidman, Angelina Jolie, Sarah Jessica Parker), comedians (Jon Stewart, Jay Leno, Jerry Seinfeld), sports stars (Randy Johnson, David Wells, LeBron James), British royalty (Prince Charles, Prince William), the two surviving Beatles, a disproportionate number of Nobel Prize winners, half of *Forbes* magazine's top ten most powerful celebrities, nearly half of baseball's Hall of Famers, an eye-popping count of top military leaders, and four of the last six United States presidents—all can claim left-handed status. [15]

Why would so many of these mysterious, doomed lefties be dominating the news? Is it revenge? Some sort of compensation? Or is there something about living in a right-handed world that gives the left-handed an edge? If so, perhaps the clumsy, gauche southpaws have something to teach the world.

The truth is that the left-handed are not a dying breed. In fact, their numbers are on the rise. The corkscrews and camcorders, insults and indignations have frustrated and inconvenienced them throughout their lives, but some of the obstacles they face—both physical and psychological—may actually give lefties an upper hand in life.

Recent advances in genetics, neuroscience, and anthropology have yielded new evidence that left-handedness is not an unintended consequence. It is one of Darwin's variations, a trait important enough to have survived at relatively stable rates since early man, perhaps even before. So it must offer advantages. What are they?

The Left Stuff takes an outside-in look at the many forces that shape—and are shaped by—handedness. It sheds light on the complexities of left-handedness, and the extent to which culture, human psychology, neuroscience, genetics, and evolution can explain what it means to be left-handed. Translating the latest science, *The Left Stuff* seeks to uncover the truth behind the many mysteries and myths that surround the embattled trait: if the left-handed do suffer so many problems, why have they survived? And if the world was designed to thwart them, why have so many of them thrived?

2

The Right-Wing Conspiracy:

The Historical Bias Against the Left Hand

Then shall he say also unto them on the left
hand, 'Depart from me, ye cursed, into everlasting
fire, prepared for the devil and his angels.'
    —MATTHEW 25:41, THE LAST JUDGMENT

In Russia you might spot them at a checkout or in a restaurant, awkwardly gripping a pen in their right hand, scribbling a sloppy name on a bill, wincing like a teenager trying to forge a parent's signature. Fearing recriminations, they learn to use their unpracticed hand in public places, or else save writing tasks for when no one is watching. Tatyana Aparshina, a saleswoman outside of Moscow, avoids writing anything in a post office or a bank, instead preparing all her documents before she arrives. Yet even the privacy of home is not always a sanctuary. Soon after she married, Tatyana was afraid to reveal her true handedness to her new husband. This was not Stalinist Russia; this was the late 1990s. But some stigmas are slow to die.

Vladimir Druzhinin, a Russian psychologist, counseled a man who wanted to convert his young son to right-handedness before putting the left-hander in charge of the family business. The man had read that the left-handed do not make good leaders—this despite the left-handedness of United States presidents who helped to end Communist rule. "The strict and rigid system of authoritarian upbringing demanded that all left-handers were forced to become right-handers," Druzhinin told reporter Will Englund of the *Baltimore Sun*. Well into the 1970s, in fact, Soviet bloc countries had strict policies against left-handedness, with teachers going so far as to tie heavy weights to the left hand to make it impossible to lift. Today left-handedness is permitted in the schools, but a prejudice endures. "Left-handers are different, and that makes them stand out, and that makes them worthy of suspicion and public comment," says Englund. For Aparshina, public comments have included, "Why haven't they taught you to write properly?" So she takes it underground. In the Russian language, doing something on the left means doing it under the table, illegally and hastily, which is what it must feel like when a lefty wants to write or eat in peace. [1]

As for Aparshina's secret, she discovered that her husband had one too. One day, a few years after they were married, he was caught peeling potatoes . . . with his left hand. She finally confessed.

Cultural biases against the left hand persist today around the world, and the force of prejudice can be powerful enough to influence the number of left-handers a society produces. Suspicions surrounding the trait may have been powerful enough over the years to limit the mating prospects of those who possessed it, leaving fewer of them among us at different points in history. In the early part of the twentieth century, for example, only two percent of Westerners were left-handed, and according to University College London professor Chris McManus, those who did possess the trait had significantly fewer children on average than their right-handed counterparts. (The trend reversed in the latter part of the twentieth century, with the presence of a left-handed parent actually resulting in more children per family, according to McManus.) [2]

Bias against the left hand is so deeply ingrained that we may hardly notice its existence today. We rarely think twice before shaking hands, taking oaths, pledging allegiance, and making the sign of the cross—exclusively with our right hands. One of the only ceremonial roles we assign the left hand is that of wedding ring holder. The Maori people of the South Pacific believed wearing the metal

THE BIBLICAL LEFT

References yoking the left to the dark side can be found throughout the Bible. In the Old and New Testaments, twenty-one verses refer to the right hand as the favorite hand, and another twenty-nine refer to the right hand of God, notes Italian neuropsychologist Franco Fabbro, who conducted an analysis of all sixty-six books of the Bible. "The superiority of the right hand also reflects God's supreme might," says Fabbro, "since the divine fire—the wisdom of God—originates from his right hand (Deuteronomy 33.2), and the most important person or friend has to sit at one's right side (Psalms 45.10)." In the Gospel of Matthew (25:31–34), when the "Son of man comes in his glory" he will separate all the nations and he will place the sheep at his right, but the goats at the left, and the former will be glorified, but the latter will be chased away into the eternal fire.

God's right hand "rescues the oppressed, punishes the enemy, gives the land to his people and administers justice" (Psalms 16:8, 20:6–7, 44:4, 48:10). In fact, the whole of creation is the work of God's right hand, according to the Bible (Isaiah 48.13); the left hand is granted only five unremarkable citations. In the book of Matthew, the Vision of Judgment depicts the King blessing those on his right, telling them they will "inherit the Kingdom," while instructing those on the left to "depart . . . cursed . . . into everlasting fire" (Matthew: 5:29, 6:3, 27:29). [8]

amulet on the ring finger protected the body's "weaker" side from temptation and other evils. Ancient Greeks, Egyptians, and Romans wore a ring on the fourth finger of the left hand to protect themselves from witchcraft. [3]

Handism is also deeply embedded in our language: according to *Webster's Dictionary, left-handed* can mean "insincere, indecisive, perhaps malevolent," as in a "left-handed compliment." In *Roget's Thesaurus,* synonyms offered for the term include oblique, insincere, clumsy, and insulting. Science texts still refer to the left-handed as *sinistrals,* the noun for sinister, which means bad omen, evil, perverse, unlucky, or unfavorable. Sinis"trals" right-handed counterparts, the dextrals, are simply skillful. The word *left* itself is derived from the Celtic *lyft,* which means weak, while right comes from the Latin *rectus,* meaning "erect" and "just." In England, a "left-handed wife" is a mistress, a "left-handed marriage" is one of social unequals, and the "daughter of the left hand" is a child born out of wedlock. Catholics are taunted by Protestants as left-footers, the off-kilter are out in left field, and bad dancers have two left feet. [4]

Leftism and Religion

During the early 1900s, a French scholar by the name of Robert Hertz decided to investigate the origins of the right hand's cultural supremacy. Studying religious doctrines, indigenous peoples, and ancient tribal rituals, he found a surprisingly universal theme. "Human hands are the inevitable symbols of all the fundamental dualisms underlying religious thought: good and evil, sacred and profane, the divine and the demonic," Hertz wrote in a 1909 essay. [5]

In fact, stigmatization of the left hand and reification of the right can be found across religious texts, from Judeo-Christianity to Buddhism. Catholic rites such as taking Communion, crossing oneself, and marking benedictions are all required to be performed with the right hand. Even in the Greek Orthodox faith, where the cross is signed in the opposite direction of the Catholic ritual, it still must be performed with the right hand. To cross oneself left-handed is considered blasphemous. [6]

From the left-lurking serpent to the line of the damned on Judgment Day, the port side is not a place you want to be. A biblical reference to the devils lurking on the left side of God is behind the modern-day custom of throwing a pinch of spilled salt over the left shoulder, and the devil is consistently depicted in religious iconography as holding a pitchfork in his left hand.

Eastern religions are not necessarily more enlightened when it comes to the left. In Buddhism, the path to Nirvana is forked: the left-hand path is to be rejected while the one on the right is to be followed, ultimately leading to nirvana. In Tantra, Indian traditions with roots in both Hinduism and Buddhism, the right-hand path consists of traditional Hindu practices such as asceticism, while the left-hand path includes ritual practices that go against the grain of mainstream Hinuism, including sexual rituals, consumption of alcohol and other intoxicants, animal sacrifice, and flesh-eating. Though the two paths are viewed by Tantrists as equally valid approaches to enlightenment, the left hand path is considered to be the faster and more dangerous of the two, not suitable for all practitioners. Hindu pilgrims circle a central temple in a clockwise or "rightward" fashion, following a ritual that Krishna is believed to have performed at a sacred mountain, and throughout India and Nepal, strangers will crisscross a path to ensure that their right sides are facing each other. [7]

For their part, the early Christians simply reflected the prevailing wisdom of the larger Greek and Roman traditions, which had established rituals against the left side. In *Life of Pythagoras*, Jamblichus wrote that the master recommended his disciples "enter holy places by the right, which is . . . divine, and leave them by the left, the symbol of . . . dissolution." Aristotle would later observe the same pattern, saying, "They call good what is on the right, above and in front, and bad what is on the left, below and behind." [9]

Plato, one of the most influential philosophers of Western civilization, linked the leftward route with evil, despite being conflicted about the merits of left-handedness in his more practical writings. In the myth of Er the Pamphylian, Plato writes about the soul leaving the body and entering the gates of Heaven, where it is met by judges sitting between two openings, waiting to pronounce their sentences. "They order the just to take the right-hand road which leads to Heaven, after having attached to them, in front, a decree setting forth their judgment; but they order criminals to take the left-hand path leading downwards, they also carrying, but attached behind, a document on which is written all their deeds." [10]

The Greeks strongly adhered to the tradition of promoting right-handedness, as did early Romans, and both civilizations developed left-to-right alphabets, which favor the "pulled pen" of a right-hander. Ironically, the perception of the left-handed as untrustworthy may have originated with a few clever lefties who learned to manipulate right-handed customs. The Roman ritual of touching right hands—the precursor to the modern-day handshake— was originally intended to demonstrate that one was weaponless. It was allegedly promoted by the left-handed Julius Caesar, who could use it to conceal a weapon in his dominant hand. Interestingly, this prejudice against the trait is not unique to highly organized, literate societies. In fact, some of the strongest militancy against the left hand can be found in the rites and practices of the most isolated cultures—even those without formal religion, written codes, or even handwriting. In Robert Hertz's studies of indigenous people across the continents, from the South Pacific to Africa to the Dutch East Indies, he found that all of them held deeply entrenched biases against everything "sinister"-sided. The Maori of the South Pacific were known for enforcing some of the most severe sanctions against the left-handed. They believed the right side to be the "side of life" whereas the left represented the side of death. Women were required to weave ceremonial cloth with the right hand, as the left hand would profane and curse the cloth. If they used their left hand to weave, they could be killed. A continent away, the Zulus of southern Africa poured boiling water into a hole then placed a child's left hand in it, with the intention of scalding it so that it could not be used. African tribes along the Niger River forbade women to prepare food with the left hand, and accused them of poisoning or sorcery if they dared to disobey. In the Dutch East Indies, a child's left arm was completely bound. (On these islands, one indicator of a well-brought-up child was a left hand incapable of independent action.) [11]

Hertz also found widespread practice of "ark of covenant" rituals, in which communities formed circles around altars that represented the place where the gods descended. Consistently, worshippers would turn their right shoulder toward the altar. "The right is the inside, the finite, assured well-being, and peace; the left is the outside, the infinite, hostile, and the perpetual menace of evil."

All of this alignment of the left with darkness may not have been entirely arbitrary. Dating back to the very dawn of agrarian civilization, before the Greeks and Romans, our ancestors were fixated on the sun. Some of the earliest religious rites and observations were connected to movements of "the big ball of light," and in the Northern Hemisphere, where most of early humanity lived, the sun moved from east to west—or rightward—across the sky. The ancient Celts practiced rites and rituals that moved deliberately counter to the left-to-right movement of the sun across the sky. They even had a word for it, *widdershins*, an adaptation of the German word *wiederschein*, which means "against the sun." If you walked around a bonfire counterclockwise, or widdershins, you were suspected of harboring a penchant for darkness, moving, as you were, against the sun. [12]

Similar rites were practiced by Hindu and Eastern Orthodox religions, including a form of the "ark of covenant" ritual, in which a person or object is consecrated as people surround it in a circle and move from left to right, like the sun, with their right side turned inward. "In this way they pour upon whatever is enclosed in the sacred circle the holy and beneficent virtue which emanates from the right side," says Hertz. "The contrary movement and position, in similar circumstances, would be sacrilegious and unlucky." [13]

These "contrary" or leftward movements could even be used as grounds for execution. In 1604, King James I of England signed a law prohibiting the practice of witchcraft and "demon worship," which led to the execution of suspected witches across England. Left-handedness, along with circling a bonfire widdershins, was used as evidence of demonic possession, as were large moles or birthmarks on the left side of the body. These witch hunts soon spread to the United States, where a preference for the left hand was used as evidence at the Salem Witch Trials in 1692, ultimately leading to the burning of Mary Barker:

> *The Examination and Confession of Mary Barker of Andover. After severall questions propounded and negative answ'rs Returned she at last acknowledged that Goody Johnson made her a witch, And sometime last sumer she made a red mark in the devils book with the fore finger of her Left hand, And the Devil would have her hurt Martha Sprague, Rose Foster and Abigail Martin which she did upon saturday and sabath Day last, she said she was not above a quarter of an hour in comeing down from Andover to Salem.* [14]

The "Female" Left

While left-handedness may be more prevalent among men, it is perhaps not surprising that both religious doctrine and mythology have tended to feminize the "lesser" side of the body. The Indian god Shiva is depicted with the right side of a man's body and the left side of a woman's body. In yoga, which has its origins in Hinduism, the left side of the body is considered the female side and the right is viewed as the male side. The goal of Hatha yoga, the most commonly practiced form of yoga in the Western world, is to harmonize this left-right, female-male duality. This is done by moving "vital energy" through the two channels where this energy is believed to flow, one starting from the "masculine" right nostril, which is linked to the sun, and the other from the "feminine" left nostril, which is related to the moon. In Kabbalah, the ancient Jewish mystical tradition, the left-hand pillar of the Tree of Life is considered the feminine pillar, embodying the characteristics of "Understanding, Judgment (or Power), and Glory." The right-hand pillar is considered to be the masculine pillar, representing the characteristics of "Wisdom, Mercy, and Endurance." [15]

A connection between gender and sidedness can also be seen throughout the Judeo-Christian tradition, where the right side of the body represents "the first stage of Creation, daytime, consciousness, Adam, Man and active power." The left represents "the second stage of Creation, Earth, matter, right, Eve, Women and receptivity." Hertz speculates that "undoubtedly God took one of Adam's left ribs to create Eve, for one and the same essence characterizes woman and the left side of the body." [16]

One of the earliest examples of the "female left" comes from the ancient Celts, whose priests and priestesses worshipped nature and viewed the left-hand side—along with the female and the moon—as sacred, the source of all life. With the spread of Christianity, however, the mystical and sacred view of the feminine role was recast as "weak" at best and "evil" at worst, and the left side was recast along with it. Readers of *The Da Vinci Code* were reminded that the Catholic Inquisition, through its publication of a text known as *The Witches' Hammer*, warned the world of "the dangers of freethinking women," including any who were "suspiciously attuned to the natural world." "Not even the feminine association with the left-hand side could escape the Church's defamation," author Dan Brown writes in the best-selling thriller. [17]

The linking of the left to the female and the weak likely predates the Catholic Church, however. In ancient Greece, the Pythagorean

Table of Opposites, which mixed number magic and mythology, designated the right side as the masculine side, connected to force, goodness, and, of course, men. In the sixth century B.C., the Greek philosopher Parmenides wrote that embryos lying on the left side of the womb were female, while those lying on the right were male. His contemporary, Anaxagoras, believed that semen from the right testes generated sons, while the semen of the left testes produced daughters. For centuries, fathers-to-be went so far as to tie off one testicle in an attempt to choose the sex of their offspring. [18]

In Hertz's study of nineteenth-century cultures, he also uncovered a gender relationship to handedness. Certain African tribes considered the right hand the strong "male" hand: good, lively, and designated to offer food and make presents. The left hand was "feeble, feminine, wicked, and deathful" and was used to take things away. The right side was also where high-ranking dignitaries stood, whereas plebes were placed on the left. The Waluwanga tribe of Australia used two sticks to mark the beat during ceremonies. "One is called the man and is held in the right hand, while the other, the woman, is held in the left," wrote Hertz. "Naturally, it is always the 'man' which strikes and the 'woman' which receives the blows; the right which acts, the left which submits." [19]

The Gauche and "Profane" Left

In addition to the seemingly arbitrary religious and gender associations tied to the left hand, "right centrism" also has a few practical roots. The weak, passive, and feminine view of the left hand was based in part on the fact that for the vast majority of people throughout recorded history, the right hand was dominant, and it became associated with whatever was strong and active. As the more dexterous for most people, it also became the official hand-to-mouth operative. So with health and sanitation in mind, particularly in places where early toilet paper options were limited, acts of personal hygiene and eating were separated—one hand was designated for feeding and the other for the "profane." Inevitably, this led to discrimination against people who ate, gestured, or wrote with their "unsanitary" hand. In his 1836 book, *An Account of the Manners and Customs of the Modern Egyptians*, author E.W. Lane wrote that most Arabs forbid left-handed food handling, except when the right hand has been maimed. "It is a rule with the Muslims to honor the right hand above the left: to use the right hand for all honourable purposes, and the left for actions which, though necessary, are unclean," said Lane. In most Islamic countries today, the right hand is used to touch parts of the body above the waist, and the left hand is reserved strictly for below-the-belt functions. [20]

During the 1991 Persian Gulf War, in the middle of a Saudi desert, General Norman Schwarzkopf found himself the guest of his Saudi allies at a banquet, laid with communal plates of food—and no utensils. "You reach into the plate and wad the food into a ball," Schwarzkopf recounted to *Smithsonian* magazine. "But I would never quite get my ball wadded up enough, and would dribble rice down my front as I tried to get it down my mouth." The five-star general's right hand, unlike his troops, refused to follow his commands. Schwarzkopf is a lefty, but he knew better than to use his skilled hand in front of the Saudis. Fighting the urge to switch hands and fearful of insulting his hosts, the general decided he'd better take action. "I just stuck my left hand under my rear." Although the Saudis might have wondered about the maladroit American, a diplomatic crisis was averted. [21]

The Delinquent and Loony Lefty

Compounding the religious and cultural fears of the left were the speculative reports from early scientists that inevitably codified and justified the mistrust of the trait. In 1876, an Italian psychiatrist by the name of Cesare Lombroso published *The Delinquent Male*, in which he claimed that men with narrow foreheads, protruding ears, and a predilection for the left hand were "psychological degenerates prone to crimes of violence." Four years later, Billy the Kid would be photographed wearing his holster on his left hip, the basis for the legend of the film *The Left-Handed Gun*. (The photograph was actually a reversed image, according to authors Leigh Rutledge and Richard Donley, so the holster would have been on the right. However, witnesses claim William Bonney was a sharpshooter with both hands, sometimes simultaneously.) Around the same time, English thieves refused to work with left-handed "pick lockers," for fear that they "brought bad luck and were invariably arrested." Perhaps the delinquent lefty theory had more to do with the left-hander's inadequacy cracking locks and safes designed for a righty than any penchant for criminal acts. In other words, the clumsy lefties were simply more likely to get caught. [22]

When they weren't being depicted as depraved criminals, the left-handed were more likely to be seen as outright insane. It's difficult to say where the archetype began, but the "crazy lefty" has been a stock character in the lefty lore. The pitching mound, in particular, has provided a lively stage for the loony lefty. It dates back as far as Rube Waddell, according to Bob Richardson of the *Boston Globe*. A Hall of Famer who hit his peak in the late nineteenth century, the madcap Waddell played marbles under the stands between innings, bragged of chasing fire engines and wrestling alligators off season, and led parades on game days—often forgetting when he was scheduled to pitch. Chicago White Sox legend Nick Altrock developed a shadow-boxing pantomime in the coach's box that eventually turned into an entire shtick, his role as a "baseball clown," says Richardson. Then there was Bill "Spaceman" Lee, the Red Sox pitcher from the 1970s who bragged that he meditated, read philosophy, and sprinkled marijuana on his pancakes. A few modern-day representatives have added their off-kilter off-the-mound marks, among them Ross Perot, Michael Richards (the actor who played "Kramer" on *Seinfeld*), Jim Carrey, Will Ferrell, and Angelina Jolie. [23]

The theory that left-handers suffer from a "syndrome" has only added to the myth of the loony lefty. In 1985, when the trait was being linked to a host of maladies and disorders, schizophrenia was added to the mix. Then in 1999, a study out of the University of Auckland, New Zealand, found a link between mixed-handedness and those likely to suffer from delusions and paranoia—a group at an increased risk of developing full-fledged schizophrenia. Michael Corballis and Kylie Barnett found that ambidextrals were particularly likely to be engaging in "magical ideation," a belief in "forms of causal concepts that by convention are invalid." This could include anything from superstitions and horoscopes to ESP, evil spirits, aliens, and previous lives—ideas adhered to by lefty Shirley MacLaine, for example. Acknowledging that cultural influences play the biggest role in shaping belief systems, the researchers nevertheless claimed a systematic relationship between the less-than-conventional and the less-than-specialized hand preferences. (Those with strong handedness, either to the right or left, were the least likely to believe in all things "magical," according to their study, yet it's worth noting that very few lefties are "strong left-handers.") [24]

A significant number of neuroscientists and psychiatrists dismiss any link between schizophrenia and handedness outright, claiming that the disease is far too complex. "Many neural differences have been found in those with schizophrenia, and there are several different subtypes of the disease," explains Lauren J. Harris, a professor of psychology at Michigan State University. One study found that left-handed men actually scored lower on a list of "psychoticism" indicators than did right-handed men. The conflicting findings may come down to how handedness is measured, as we will see in the coming chapter. "Sticking left-handers in the same category as those who are not conventionally dominant can produce spurious kinds of relationships," adds Harris. [25]

Yet there is no doubt that myths and stereotypes have attached themselves to the left-handed, and like many stereotypes, they can become self-fulfilling. The trait may even grant those who possess it with the unspoken permission to be different.

"There's an interesting body of research in social psychology which finds that people define themselves based on qualities that make them stand out from the norm," explains Harris. "Here in Michigan, for example, you will not have people identify themselves as 'someone having brown hair.' It's too common. But you might find that people identify themselves as left-handed. It is something that sets people apart."

"In my case, my left-handedness was routinely mentioned by my mother when I was a child. At a restaurant, she would invariably say 'Lauren has to sit over there, he's left-handed.' It wasn't a question of being diminished, more a case of being treated as 'special.' When I was in school in the 1940s, we had inkwells on our desks, and they were always on the right side, so being left-handed was an obvious difference. It became part of self-identification—a sense of oneself as being somewhat different." [26]

Left in the Twenty-First Century: A Lingering Prejudice

Despite advances in science, left-handedness remained a trait to be suppressed well into the twentieth century. In the early 1900s, American educator A. N. Palmer held symposiums across the United States instructing parents and teachers to force left-handed children to write with their right hands, arguing that it was important for children "to learn the value of conformity." Teaching manuals published throughout the first half of the century urged educators to train all students to be right-handed, and some went so far as to recommend tying the left hand to the back of a chair to prevent its use. During the 1960s and 1970s, teachers in and out of Catholic schools still whacked the knuckles of students who used their left hands. "I grew up in a culture down south where folks thought that people who were left-handed were somehow stained by Satan," recalls E. R. Shipp, the Pulitzer Prize–winning columnist for the *New York Daily News.* "They tried their darnedest to make me right-handed. And others before and after me. But, thank God, I'm still left-handed." [27]

Northerners did not have it much easier. In Montreal's Catholic school system in the 1960s, left-handedness was classified as an "objective disorder," recalls Derek Evans, the former deputy secretary-general of Amnesty International. "This meant that school authorities were allowed to deal with it by using discriminatory, sometimes brutal and abusive, methods." Evans attributes his lifelong commitment to human rights to his childhood experience as a left-hander. When he was a kindergartner, Evans's parents challenged the school's policies. "My parents were quite ordinary people, but for some reason they stood up to the system and insisted that I should be allowed to be the person I am. They protected me from the abuse, and they won. I was only five years old, but I learned a lot from that experience—about how brutal things can be for persons who fail to meet arbitrary standards of what is considered 'normal' and how ordinary people can sometimes bring about change by saying 'no' to the powers that be." [28]

Across the Atlantic in postwar England, treatment was not much better. As Sir Paul McCartney recalls, "When I was a kid I seemed to do everything back to front. I used to write backwards, and every time the masters at my school looked at my book, they used to throw little fits. . . . I do everything with my left hand, and no matter how hard I try I can't alter the habit." The world of music is grateful for the maladapted maladroit. When he first auditioned for the Beatles, McCartney was given a right-handed guitar. He turned it upside down and began to play in reverse. ("I remember John looking at me, like, wow, you know, this guy's got something here," McCartney reminisced to BBC Radio in 1999.) [29]

Despite the constant roadblocks, left-handedness as a whole began to gain greater acceptance in North America and Western Europe beginning in the 1940s, when the progressive education movement lead by John Dewey started to take hold, promoting a more democratic and less conforming approach to teaching methods. This open-mindedness continued in the wake of World War II, when a greater tolerance for individual differences sprung up in the wake of totalitarianism, helping the left-hander's cause. In 1946, maverick president Harry Truman became the first left-hander to throw the presidential "first pitch," which helped raise the profile and status of the trait. The rising celebrity of Ty Cobb and the greatest legend of America's pastime, Babe Ruth, also helped spawn a new appreciation for the trait (not to mention the manufacturing of more mitts for the short-shrifted southpaws). [30]

Throughout the latter half of the century, systematic pressures against left-handedness lifted in North America and Western Europe as parents became more permissive. Dr. Spock's best-selling guide to parenting and child care, which had advised parents to encourage right-handedness when it was first released in the 1940s, revised its counsel in later editions. By the 1960s, Dr. Spock was encouraging parents to adopt a more laissez-faire approach to their child's handedness. [31]

Despite these advances, biases against the trait persist, in Western countries and around the world, especially in the wake of the latest fears surrounding premature death rates, health risks, and the so-called syndrome of disorders. The bias is strong enough to revive pressures on young children to switch sides, which can lead to developmental problems in early childhood, given the intrinsic link between handedness and the brain, as we shall see. [32]

Prejudice may even be powerful enough to influence the number of lefties a country produces, or at least the numbers of those who can be identified. In cross-cultural studies conducted by social scientists, the percentage of left-handers in a society can serve as a barometer for that society's tolerance for difference. Contemporary schools in China, Korea, Taiwan, and many Muslim countries systematically suppress left-handedness through physical constraints or rebukes. In contrast, as Japan has lifted sanctions against the left-handed after World War II, their numbers have increased sharply. [33]

Under Communist rule, Soviet bloc countries had strict policies against left-handedness that were enforced by educators well into the 1970s. Decades later, the stigma surrounding the trait has been slow to die. Even in the supposedly enlightened United States, pressures against left-handedness persist. A 1990 study found that more than half (fifty-five percent) of self-classified left-handers reported an attempt by parents or teachers to make them use their right hand at some point in their lives, and approximately twenty-four percent of young lefties still face some pressure to switch hands. Trolling lefty websites as well as educators' and parenting forums, plenty of evidence surfaces that switching pressures—subtle and overt—still exist during the developmental years. "I try and guide my son gently, place the crayon on the right side, put the spoon in his right hand," one mother wrote in a recent Craig's List parenting forum. "I want to make sure he picks a side, and it would just be easier for him to be right-handed, since most things are designed for right-handers." Yet psychologists have found that even subtle nudging can lead young children to suffer feelings of guilt over disappointing a parent, or feelings of inadequacy brought on by their inability to conform. And at least two studies have found that those who experience pressure to switch sides become more anxious as adults. [34]

The belief in the value of coaxing a child toward right-handedness is likely to be more prevalent among parents, teachers, and other caregivers who have recently immigrated to the United States As recently as 1970, Spain, Italy, Yugoslavia, and all the Iron Curtain countries (with the exception of Czechoslovakia) made right-handed writing compulsory in schools. In Albania, left-handedness was illegal. Yet even in the absence of any adult pressures, some kids may simply feel ashamed of their difference and try to hide their left-handedness. In the late 1990s, New York child psychologist Jane Healy treated a five-year-old boy who was having trouble developing basic drawing and writing skills. His kindergarten teacher had informed the boy's parents that he was switching back and forth between his hands. This surprised them, as he always wrote with his left hand at home. During his first meeting with Healy, the boy used his right hand, insisting that he always did. It wasn't until she told him that she herself was left-handed that he felt comfortable switching. "Apparently when he realized I was left-handed, he decided it would be okay to do what was most natural. Despite the fact that his parents were both right-handed, he was not uncomfortable at home. But in school, where he felt he had to conform, he felt compelled to write the same way the teacher and most of the other students did." [35]

Cultural views on handedness can leave an imprint on the unconscious well into adulthood. A New Jersey–based company has gone so far as to offer a "Being Lefthanded Phobia Clinic" for people suffering from "sinistrophobia," which they define as a "surprisingly common" malady that includes a "persistent, abnormal, and unwarranted fear of things to the left or left-handed." According to the clinic's website, the condition is often manifested by panic attacks and even avoidance of individuals who are left-handed. While the clinic promises it can cure *sinistrophobia* in "two to three hours," some deeper psychoanalysis might be in order. In 1969, University of California Santa Cruz professor William Domhoff collected case histories from psychoanalysts and documented the extent to which the patients, and in some cases, the analysts themselves, internalized the cultural bias against left-handedness. In one case, a patient described his guilt over doing things left-handed in his dreams, in what he described as "the wrong way." Interpreting a dream in which a car hit an obstruction and swerved to the left, a psychoanalyst explained that "the left side of this patient, as is common, represents his weaker, and thus feminine side." When a young female patient saw herself as two people, a "right and a left Rose," her analyst wrote, "it appeared that the right one was her good self, while the left was a delinquent and utterly wicked child who could not be tolerated by the better self." Domhoff speculated that the "dichotomous thinking" attached to handedness may have increased the tendency for right-handedness, investing it with the preferred side of the "good-bad, active-passive and potent-castrated polarities that pervade so much of our thinking." [36]

* * *

And yet the trait has endured. Despite a history overflowing with knuckle-rapping nuns and ill-conceived teaching manuals, left-handedness has sneaked by the kindergarten teachers of the world, resisting persistent pressures from all directions. Their numbers may have dwindled, but in almost every culture, southpaws remain a fixed and ever-present portion of the population. Despite what religion, cultural norms, and educational approaches have preached, the left-handed are still here. And for some very good reasons.

3

Who Are the Left-Handed?

Deconstructing Lefty

Maybe it makes sense to call the conservatives the people on the right, because there's a certain homogeneity to being right-handed.
They tend to also be right-eye dominant and right-footed. Left-handed people are more likely to have mixed dominance.

—Jeannette Ward, professor emeritus, the University of Memphis

Throughout most of his life, Bill Gates did everything "right" but write. For that task, the richest man in the world uses his left hand. During a time in college, he gave his left hand a break and started taking notes with his untrained right hand, "just for the heck of it." But today, Microsoft's chairman and chief software architect has returned to his left-scribbling ways. Yet while his company might sell a left-handed mouse, he prefers to use the right-handed model, clicking and dragging strictly from the starboard side of the keyboard. [1]

Can lefties claim Bill Gates as their own? What about Babe Ruth, Martina Navratilova, and the roughly six percent of the modern-day population who write *right* while performing most other tasks *left*? How about lefty-swinging golf champs Phil Mickelson, Mike Weir, and Steve Flesch, who perform all other tasks with their right hands? Should they be sharing a mantle with right-handed golfers Arnold Palmer and Ben Hogan, who perform most nongolfing activities left-handed? Where do we put the right-eyed, left-smiling, right-footed, left-eared who happen to write with their left hand?

The answer to these questions is at the root of many of the mysteries and much of the confusion that surrounds handedness. Just who can claim the lefty identity? And how many of them are there? Although the historical bias against the left hand has always been clear, the actual definition of what it means to be left-handed has not. Any objective look at handedness is complicated by the fact that a classification of people into Left or Right isn't so black and white. Because so few people use only one hand to do everything, handedness is more of a continuum than an either/or identity.

The laity does not necessarily see it that way. On average, people in self-reporting questionnaires identify themselves as left-handed just over ten percent of the time. Yet studies that compare self-described handedness with observed handedness show very different results. "People by and large are using their left hand much more than they are consciously aware of," explains Jeannette Ward, a professor of comparative physiology at the University of Memphis. In fact, one study found that thirty-five percent of people use their left hands for certain revealing tasks such as throwing, eating with a spoon, drawing an object, striking a match, or sweeping with a broom (left hand on top). "There's the expression the right hand doesn't know what the left hand is doing," says Ward. "Well, a lot of people don't know what their left hand is doing. It's not even on their radar screen." [2]

But does using your left hand—unwittingly or selectively—make you a closet lefty? Psychologists and neuroscientists have wrestled over the definition of handedness for more than a century, and for good reason. The brain and the body are cross-wired, which means that the right side of the body is controlled by the left side of the brain, and vice versa. In the first half of the twentieth century, brain surgeons relied on a person's hand preference to tell them which side of the head to operate on: if a person had lost motor control on the left side of his or her body, surgeons "opened the hood" of the right side of the brain.

Today, the notion of a left-handed "syndrome" and the studies that link the trait to disorders, dysfunctions, talents, and deficits usually fall apart when the definition of the trait—and criteria for measuring it—is changed. [3]

Scientists may never completely agree on who the "real" lefties are, in part because of the variety of combinations into which handedness (not to mention sidedness) falls. The discussion grows even more complicated when scientists factor in the various assumptions about the meaning of handedness and the extent to which it represents the organization of the brain—or even the evolution of the species. Is Bill Gates a lefty, when all he does with his port-side hand is write? How can he share a mantle with Martina Navratilova, who does everything left-handed *except* write? There's a subset of people who actually draw with one hand and write with the other. Another group *prefers* their left hand, but actually *performs* better with their right. Some people who call themselves righties might actually have stronger left hands. And what to do with the countless professional athletes who play it both ways—swinging left, throwing right, backhanding and forehanding until no one knows which paw is south?

Defining *Lefty*

In the early part of the twentieth century, lefties were simply known as the people who wrote with the Other Hand. It was simple enough to observe, and simple enough to count. Yet studies that have looked at the relationship between left-handedness and other traits turn up very different results when "writing hand" is replaced with broader criteria for handedness. [4]

Over the years, handwriting, as one of the most visible displays of hand difference, has been the prime target for right-handed proselytizers, be they nuns and disciplinarians or parents who simply believed that conforming to the right-handed world would make life easier for their kids. Even those forced to shift their allegiance to the right were usually free to do other things—throw a ball, brush their teeth, or thread a needle—with the hand that felt more natural. If handwriting were the only measure of left-handedness, Babe Ruth would be considered a righty. Ruth's reform-school teachers successfully switched his handwriting. Luckily, his throwing arm and batting swing were left alone. [5]

Does someone who was subtly coerced or abusively forced to use their right hand truly become right-handed? Ronald Reagan put forks and pens in his right hand, but as an actor, he twirled and shot fake guns with his left hand, a fine-motor skill not usually targeted for conversion. During his presidency, when he threw out the first pitch at Memorial Stadium using his "northpaw," he lost control of the ball and threw it into the stands. Perhaps he would have had better luck had he thrown lefty. Similarly, former presidential candidate and senator Bill Bradley has spent twenty-five years on the left-handed lists, the result of scholarly monitoring of his left-handed basketball skills. But he does not write or eat left-handed. If his left hand did prove to be faster and more coordinated when tested on the court and in the lab, what would that tell us? Does a superior left-handed layup betray an inner lefty? [6]

A major challenge facing researchers who set out to answer these questions is that left-handers are harder to find and recruit for research, since there are far fewer of them to begin with. Further complicating the study of southpaws is that whenever any small group is studied as a population, their bell curves become exaggerated compared to the majority. "Much smaller numbers get much larger effects," explains Ward. "There's a much higher probability of noticing left-handedness than right-handedness, given the size of the populations." If left-handers appear to pop up disproportionately in studies of everything from bedwetters to alcoholics, it may come down to two or three extra people who ended up boosting the sample size.

One of the most widely published scholars of left-handedness is Marian Annett, a psychology professor at the University of Leicester in England. According to Annett, the trait does not have a simple, agreed-upon definition. "It is because it can be defined on any of several levels of preference (from weak to strong) that confusion exists in the literature. Whether or not someone calls themselves left-handed depends on the actions which are important to them, or their questionnaire." [7]

Most researchers measure handedness in two ways. The first is to ask questions about eight to twelve different activities and let people gauge for themselves which hand they'd likely use. The second involves actually testing hand skill, observing as a person puts pegs into holes or taps his or her fingers, and measuring the difference in the accuracy and speed between the right and left hands. When it comes to the final outcomes, roughly the same numbers of lefties tend to turn up whether testing for preference or performance. They're just different subsets of lefties. While there's considerable overlap between the two groups, counting both of them would bump up the total percentage of left-handers in the population—to as many as thirty-five percent of the population. [8]

Once it's decided which test will be used to measure handedness, there is the question of how to group the outcomes. Some follow the dichotomy hypothesis, dividing people into two simple categories: the consistently right-handed and everybody else. This is based on a theory that those with even slight left-handed or ambidextral tendencies share the same type of genes—and similar patterns of brain organization—as those who are consistently left-handed.

Other neuroscientists argue that the significant difference is not the direction of handedness but the degree—those who are strongly left-handed and strongly right-handed have more in common than the "mildly" handed, or people who perform at least two out of ten important tasks with their nondominant hand. Cognitive tests and even cadaver research has turned up differences between these two groups, as we shall see. So this would mean the left-writing Bill Gates and right-swinging Arnold Palmer belong in the same group, while purist lefties and purist righties are on the same team. It gets even more complicated when scientists subdivide people further, sorting out the degree and strength of their hand preference. Marian Annett argues that the non-right-handed are comprised of additional subgroups with different gene combinations, each leading to a different brain organization, and therefore different skills. [9]

The Handedness Tests

To measure the degree of handedness, researchers rely on a popular tool called a handedness inventory, a list of several different tasks often accompanied by a scale for each that allows respondents to report which hand they use and how often or how strongly they prefer it. Each inventory is based on different research assumptions, so it can measure different underlying forces: the hand that performs faster on manual tests, the hand that one prefers to use regardless of performance, the hand that is used consistently over time, or the hand that is used consistently over a range of tasks that require fine motor skills. For example, in one inventory, someone who uses his or her right hand some of the time would end up in the same category as someone who always uses his or her left hand.

The most revealing of the inventories, created by Annett in 1970, measures for twelve tasks and allows the outcomes to be divided into eight groups, a full spectrum that ranges from consistent right-handers, to right-writers with weak left-handed preferences, all the way to consistent lefties. By breaking down people into many subsets, Annett found that "weak righties," those who perform a few tasks with their left hand, have more in common with strong left-handers than they do with their fellow right-handers who are more consistent in their leanings. Even a slight left-handed propensity may indicate the presence of "left-handed" genetic tendencies. Investigate the family tree of a tried-and-true lefty, and you are likely to find a couple of "mild righties," according to Annett. [10] (See appendix for complete questionnaire.)

A simplified version of Annett's survey, the Edinburgh Handedness Inventory, is the most common survey used today. Measuring the trait based on self-reported preferences for performing ten simple tasks (such as writing, throwing, using a spoon, using a toothbrush, striking a match, or opening the lid of a jar), the Edinburgh results are quantified into a "laterality quotient," which allows people to fall on a continuum ranging from twenty (completely right-handed) to minus twenty (completely left-handed). [11] (See Appendix.)

Examples of this testing confusion permeate the scientific literature on the definition of handedness. One graduate student in psychology consulted for this book, Karen Caldwell, represents a good case study. As an undergrad, Caldwell volunteered for several different research experiments that included handedness questionnaires, some of which left her very puzzled about her identity. "One study had me in the right-handed group, and another said I was a lefty," Caldwell recalls, noting that the only things she does left-handed are throw a ball and hold a broom handle with her left hand on top—two tasks she rarely undertakes. "Who knew that sweeping could 'out' me as a lefty?!"

Well, not necessarily, but throwing a ball left-handed is a sign that Caldwell very likely does not share the brain organization of the average right-hander. Performing one or two single-handed activities from the inventory list consistently with the left hand is enough to anoint someone left-handed, according to some scholars. [12]

Yet even something as seemingly simple as the word *consistent* can be a problem. One of the fundamental flaws to any verbal test is linguistic. Research outcomes can be affected by the language people feel comfortable using to describe themselves. Surveys that ask a person whether they "always" or "usually" use their right hand for something—say, opening a lid or using a broom—may succeed only in separating out those most willing to use very strong language, words such as *always* and *never*, from those who prefer more ambiguous language, like *usually* and *sometimes*. [13]

These limits on language have led British neuroscientist Dorothy Bishop and her colleagues to argue that the best way for measuring the strength of a person's hand preference is to observe how readily a person uses his or her nonpreferred hand in a reaching task such as grabbing for a pen on a desk. The left-handed are more likely to register objects that are in the left side of their field of vision, and are therefore more likely to reach for them. When measuring the degree of handedness, however, Bishop's research found that the most useful handedness inventory measures the pure number of activities performed by the preferred hand. [14]

The Formative Years: When We Pick Sides

There's an old wives' tale that says if a newborn's hair whorls counterclockwise, they are going to be left-handed. The old wives must've kept fairly good statistics, because a 2003 finding out of the National Cancer Institute in Maryland recently proved them right.

In a study that looked at the direction of hair growth from the crown of the head, more than ninety-five percent of adult right-handers had a clockwise coil. But in lefties and ambidextrals, fifty percent had hair that whorled counterclockwise. [15]

The spin of the first cowlick may be a long way from the squiggles of the first crayon, but predictions of handedness may present themselves in infancy. Between four to six months, most infants, future lefties and future righties, hold a rattle in their right hands longer than they do in their left. From six to nine months, the majority reach for a toy more often with their right hand than with their left—regardless of which hand the toy is closest to. Yet it's important that babies and toddlers still use their nonpreferred side. According to pediatricians, if they show a very strong preference for their right or their left sides, or rarely use one side, they could be suffering from some form of neurological damage. [16]

Only when some degree of skill is required can true "handedness"—differences in hand preference and performance—emerge. The earliest point at which we start to become specialized with our hands is at the end of the first year of life, which is roughly when brain lateralization first asserts itself. This is when most babies are capable of reaching, pointing, and performing the kinds of nuts-and-bolts tasks that are used to test adults for hand preference later in life—placing pegs in holes or tapping fingers, for example. Parents can place an object in front of their sitting infant, such that each hand has equal access, and begin to detect a pattern when it comes to which hand is used most often to reach for the object. One study also found that one-year olds are more likely to use their right hand to point than their left. This is a trend that carries over into other activities. At least two-thirds of one-year-olds turn a bolt with their right hand while holding the nut with their left—the same percentage as with adults. [17]

Despite this apparent preference, in the early stages of hand skills, toddlers tend to switch back and forth between their left and right hands. This is known as the chaotic phase. By the age of three, most kids have developed a preferred side for specific tasks, even if it means using one hand for moving pegs and another for coloring. Lefties are more likely to use different hands both across and within tasks, but even an early left-hand preference at this age does not predict that a toddler will become left-handed. Hand choice is not usually stabilized until after age four. The biggest differences in hand skill happen between three and six, when kids are trained to hold a pencil and write. By three or four, most kids are tackling a few coordinated, complex activities such as coloring, cutting with scissors, or manipulating a toy. At six years of age, most kids have made up their minds, and those previously classified as mixed-handed become right- or left-handed. Of the kids who tested mixed-handed at five, only four percent remained so at age eleven, with the majority becoming right-handed. [18]

Ambidextrals, Mixed-Handers, and the "Flip-Flop" Variable

President Gerald Ford is a particularly interesting case for handedness scholars. The thirty-eighth president does most everything right-handed while he's standing up (throwing a ball, swinging a golf club) and left-handed when he's sitting down (such as writing and eating), a not-so-rare pattern that proves to be very revealing in the coming chapters. Some might assume that Ford is ambidextrous, another label for the in-betweens, those who use their opposite hand for certain tasks. The true ambidextrals never check either/or for a task; they consistently use the left hand for certain tasks and the right hand for others. The most common ambi archetype is someone who writes or eats right-handed but tackles most other single-handed activities left-handed. These people are often "switched" lefties who were coached or coerced away from the left during childhood. [19]

Switch-hitters such as record-breaker Barry Bonds and Yankee legend Mickey Mantle are not, technically speaking, ambidextrals. Many lefties learn to swing a bat or a golf club with their right hand on top. Some would argue this is even to their advantage, allowing their stronger left arm to do the heavy pulling. The problem is that any two-handed—and especially two-armed—activities are not very significant in predicting a person's handedness, as they involve the whole upper body and tend to be very responsive to coaching and practice. A young slugger with a Wiffle-ball bat may emulate an older sibling's swing, then feel more comfortable with this early adaptation as he or she gets older. Other "power" movements that involve both hands can confuse the left-right identity. Kevin Laland, a geneticist at Cambridge University in England, explains, "If I ask you if you open a jar, which hand holds the lid, you might have to think about it, and be less certain. You might give different answers at different times, or you might want to answer sometimes left and sometimes right, depending on whether I require strength or dexterity." [20]

There are actually a large number of left-handers, up to thirty percent, who write with their left hand and throw with their right. Given the various pressures on lefties to write right-handed, it doesn't make much sense that someone who resisted pressure to switch his or her writing hand would end up throwing right. But according to Michael Peters, professor of psychology at the University of Guelph in Canada, this group may have a different brain organization when it comes to motor control, as we shall see in coming chapters.

Cognitive psychologist Stephen Christman, who has conducted several handedness studies at both the University of Washington and University of Toledo, finds that the most significant differences show up when comparing mixed-handers to strongly handed persons. To be classified as a strongly handed individual, a subject must perform nine or ten activities always with the same hand, which include write, draw, eat with a spoon, throw a ball, brush your teeth, strike a match, open a jar, hold a broom (upper hand), and use scissors. If a subject performs eight of these with one hand, performs one with the other hand, and has no preference for the tenth activity, then he or she would be classified as mixed-handed. [21]

But when it comes to focusing on the fine motor skills, those performed with only the hand (and not the whole arm), many scientists discount the mixed-handedness categories all together. "Very few people are truly ambidextrous," Peters explains. Survey responses may cast a person as mixed, but when it comes to measuring actual performance, one hand is usually weaker or less skilled than the other. In finger-tapping exercises, another useful test of hand skills where researchers measure the intervals between taps on a keyboard-like machine, the studies almost always reveal a stronger, quicker, and more skilled hand. [22]

Inevitably, the people who don't become fully specialized with either hand are the ones who interest researchers the most. Unlike ambidextrality, *ambiguously* handed people choose either their left or right hand for the *same* task. In research findings, this indecisive group tends to get buried among the non-right-handers or the ambidextrals, but they should be sorted out because "flip-flopping" on a task can be a red flag for certain neurological or developmental problems. The developmentally disabled, for example, are more likely to be inconsistent in their hand choice, particularly if they never fully learn to write. So are young children with delayed development—the subjects of many handedness studies—who also tend to struggle with speech and reading problems in the early years. "We've realized that the important variable is not classifying a person as left- or right-handed or mixed-handed, but the consistency with which they use their hands," says Robert A. Hicks, a psychology professor at San Jose State University. Hicks and his colleagues first isolated this group by looking at large handedness studies and pulling out the respondents who had checked "either/or" for six or more tasks, a small but significant group that often gets thrown in with the lefties. Once they were sifted from various findings, it became clear this subgroup was the at-risk population when it came to developmental problems or even accidents. Lumped in with other non-right-handers, these "undecideds" were skewing the data on the overall left-handed population. [23]

The "conflicted" may also explain how certain studies have linked left-handedness to schizophrenia while others have not. According to Peters, an editor of the journal *Laterality*, the research suggests that schizophrenics "are as right-handed as anybody else" when it comes to performance and writing hand. Yet they are less consistent in their preference choices and in their responses to questionnaires. "When you and I fill out a preference questionnaire, we are more concerned with being consistent than a person with schizophrenia, and will therefore make slightly more categorical choices. That is all it takes. If schizophrenics are florid and out of it, one may not be able to put much faith in their choices altogether." [24]

The Common Denominator

While there may never be a consensus on who qualifies as an official left-hander, researchers do tend to gather around a common denominator definition, which goes something like this: a left-hander is a person who uses his or her left hand for activities that require a lot of practice and fine motor skills, such as writing, or involve the coordination of large muscle groups to carry out smooth actions, such as throwing a ball. These are also activities that use a lot of neurons in the brain and require a tightly concentrated and "specialized" neurological wiring. If a person tackles several specialized skills—such as eating, using a razor, combing hair, or cutting bread—with his or her left hand, then he or she is left-handed. And if the person embarks on the larger "gross" motor skills such as hitting a tennis ball, throwing a ball, and hammering a nail with the left hand, then he or she is strongly left-handed. But all it takes is one of the big three—writing or eating or throwing—to get you in the club.

A Raise of Hands:
A Count of Lefties

Based on studies that combine handedness surveys, approximately twelve percent of Americans and Western Europeans are left-handed, with breakouts varying depending on sex, age, and cultural background. While these rates are relatively consistent in Western countries, they vary considerably around the world. The more permissive a culture is about individual differences in general, the more left-handers appear. The scarcest populations of lefties are found in Asian, Muslim, and Latin American countries, "formal cultures" that stress conformity and fixed sex roles. (North American cultures are classified as nonformal.) These countries continue to systematically switch young kids to right-handedness, particularly when it comes to writing and eating. Korea, Japan, and Taiwan churn out the lowest levels of lefties, with estimates averaging two, three, and five percent respectively. In a cross-cultural study of self-perceived handedness, Belgians came in first at 15.7 percent, while Nigerians were the least likely to report left-handedness, at only 4.9 percent. And while most of England counts the left-handed as eleven to twelve percent of its citizens, lefties are scarcer in the Celtic isles of Wales and Scotland, representing a mere eight percent of their populations. This could be attributed to more traditional teaching methods in the highlands, or it may be that the homegrown lefties were more prone to wanderlust, and headed to the motherland. Studies have found that nomadic people tend to be more left-handed. [25]

Although cultural values clearly influence the numbers of lefties a country produces (or at least the numbers reported), there are other variables at work. Japan has very few left-handed writers, but one reason for this is the complexity of writing in logographic script—the familiar Chinese and Japanese characters. "There is great emphasis on the direction and quality of brush strokes, and it is of great advantage if both teacher and student share the same handedness," explains Peters. Technical aspects of writing may have played a role in the relatively recent increase in left-handed writing in Western countries as well. Before ballpoints and lead pencils, people had to write with course nibs on rough paper, so it helped if you could pull the nib across the paper, adds Peters. If it was pushed with the hand of a lefty, it could get stuck or create ink splotches. [26]

To get a sense of the true proportion of left-handedness in a country, scholars look at activities other than writing. For example, in a 1983 study of Japanese boys, the number who used their left hand to "whittle with a penknife" was close to nine percent. Yet with ongoing pressures to conform, students will increasingly opt for the right hand as they get older, even for activities for which the left hand was preferred initially. This is a trend that occurs throughout Asian countries. Often, those who might have become left-handed in a different culture are classified as right-handers, according to a study conducted by Michigan State professor Lauren Harris and his Korean colleague Yeonwook Kang. They found that while a left-hand preference is higher for nonwriting activities than for writing, overall left-handedness among Koreans was still much lower than in Western countries, 4.2 percent compared to an average twelve percent. "Public acts such as writing and handling food are severely restricted in Korea, more so than in other Asian countries," explains Harris, adding that the offering and acceptance of gifts is also conducted exclusively with the right hand. "The supposition is that a gentler kind of pressure is asserted on other acts, but it's possible that once you're trained to be right-handed for writing and handling food, this training will generalize to other tasks that are not specifically targeted as part of social training practices." [27]

In the modern-day United States and Canada, where pressures to conform have weakened and writing implements have improved, the numbers average out to approximately thirteen percent of men and just under eleven percent of women. Those under thirty are more than twice as likely to write with their left hand than are those over sixty-five (fifteen percent versus six percent), a gap attributed in large part to the systematic switching of left-handers in the first half of the twentieth century. Ten percent of people in the over-sixty category claim their handwriting was successfully switched before or during primary school. [28]

Though Americans no longer face institutional pressures against lefty handwriting, indirect pressure still exists. Recent surveys have found that nearly half of younger-generation left-handers (forty-six to forty-nine percent), those born after the baby boom generation, experienced some pressure to convert their hand preference. Many ethnic and religious populations within the United States are still making more overt attempts to shift from left to right, leaving more than six percent of people aged eighteen to thirty to say that the hand they write with was successfully switched when they were kids. Because handwriting is typically the focus of shifting, other activities—throwing, teeth brushing, even drawing—tend to be left alone. Tallying it all up, with schoolchildren and those with switched handwriting, the estimates for left-handedness come to approximately twelve to fifteen percent in the United States today. [29]

The numbers get significantly larger when we include the non-right-handed, which many studies do. Annett and several of her colleagues argue that significant differences in brain organization and cognition can be found when separating the strongly right-handed from mixed-handers. They estimate that as many as thirty to thirty-five percent of the U.S. and U.K. populations qualify as non-right-handed, which means they perform at least two of the ten key tasks with their left hands. [30]

Both Sides Now

Because handedness is a motor skill and falls under the jurisdiction of the brain, it is part of our larger human tendency to choose sides, or lateralize. Though each side of the body is controlled by the opposite side of the brain, our sides specialize in different activities, a process called lateralization. One foot will consistently lead us over a puddle, the same eye will gaze into a telescope or camera, and one ear will eavesdrop on an argument through a closed door. Approximately eighty percent of human beings favor their right foot, roughly seventy percent favor their right eye, and sixty percent favor their right ear. [31]

While these preferences are, like handedness, connected to the division of the brain into two hemispheres, their relationship to handedness is not so simple. Today, scientists have studied every combination of sidedness and dominance, and the left-handed are least likely to be consistent in their "alignment" of preferences. Approximately forty percent of lefties are right-eye dominant, and nearly half of them prefer their right feet. This sort of "cross dominance" may seem like a liability, but it can actually serve as an advantage, and not just when it comes to playing Twister. "A left-handed batter with right eye dominance has an advantage when it comes to monitoring a pitch," explains Glenn Fleisig at the American Sports Medicine Institute in Birmingham, Alabama. And perhaps left-handed golfers (with otherwise right-handed tendencies) have the right-eye dominance edge when it comes to aiming and visualizing drives and putts. [32]

Eye dominance often reveals itself during one-eyed tasks such as looking into a camera or a telescope or sighting down the barrel of a gun. But the dominant eye is more than just a preference, it reflects the eye that dominates our aim and consolidates our vision—preventing us from having double vision. It is determined fairly early in life, showing up in infants as early as the first year, and often it is not the eye that performs better in eye exams. Even when it loses visual ability, the dominant eye usually remains dominant. Information received from the dominant eye is processed faster than cues sent to the nondominant eye, and there is some evidence that the colors it sees appear richer. As we shall see in the coming chapters, the split role of the two eyes can shed some light on the processing differences between left- and right-handers. [33]

Determining eye dominance is fairly simple: with both eyes open, place an index finger out at arm's length so that it covers a distant object. Close each eye separately. The eye that, when closed, results in your finger's "moving" away from the object is your dominant eye. For more than seventy percent of right-handers and approximately forty percent of left-handers, the right eye is dominant. Without it, their aim is off. Eye dominance even shapes our sense of direction and orientation to the world. Asked to locate a point "straight ahead" of them, most people pick a point that is shifted toward the side of their dominant eye. [34]

Preference for one ear is more muddled, as the ear-hand connection is much weaker. Only sixty-three percent of the population has matching ear and hand dominance, according to Stanley Coren. Tests for ear dominance often involve use of specialized techniques, administered by audiologists or psychologists. Yet Coren and his colleagues tested a large sample of subjects based on preferences and found that fifty-nine percent preferred their right ear to hear a sound through a door or to press against a clock. [35]

"The Left Stuff"

To understand the relevance of left-handedness as a human difference that has persisted in mankind throughout recorded history, the trait needs to be looked at more broadly. After all, it has thwarted worldwide and historic cultural pressures, in rates that appear to have ebbed and flowed with social tides. Yet while left-handedness may not have made itself consistently visible, it has managed to survive in the species at fairly stable rates. To fully grasp why, we need to look at what the trait tells us about the organization of the brain. Though our dominant hand is connected to how our brain develops, we don't always end up using our dominant hand, meaning that the only way to truly know which hand we're destined to prefer is to take a look inside our brains. Failing that, a few indicators can tell us something about where we fall on the hand-brain spectrum: lefties can be people who perform one or two single-handed activities from the handedness lists, be it brushing their teeth, threading a needle, dealing cards, opening the lid of a jar, throwing a ball, eating, or writing. Even more intriguing, they may check none of these activities. In fact, as far as they know, they are pure-bred, consistent right-handers—so far in the lefty closet they don't even know they're in it. But deep inside, they are wired differently from the right-handed majority. For this population, the presence of a biological sibling, parent, or child who is overtly left-handed could be a clue to their innate left-handed predisposition. We'll step inside this "predisposition" in forthcoming chapters.

4

Lefty on the Inside:
Handedness and the Brain

*"The right hemisphere is a left-wing revolutionary that
generates paradigm shifts. The left hemisphere is a diehard
conservative that maintains the status quo."*
—V. S. RAMACHANDRAN AND SANDRA
BLAKELEE, PHANTOMS OF THE BRAIN

D r. Joy Hirsch, chief of Columbia University's functional Magnetic
Resonance imaging (fMRI) Research Center, is a mirthful brunette
in bangs and a deep blue jean jacket. I first visit her at her brain-imaging
laboratory, a sleek facility with NASA-sized computers surrounded by inky
images of lumpy, snake-coiled brains. A plate-glass window separates a
control room from its subject: an enormous white labyrinth of a machine,
big enough to house a chocolate factory. At its core is a dark tunnel with a
flatbed that looks as if it could launch small rockets. "That's where you'll
be lying," Dr. Hirsch says, pointing to the narrow channel.

I have volunteered to have my head examined, a full brain dissection, during which I will be fully awake. The lab assistants have assured me that it will be entirely painless and utterly harmless. There will be no knives or incisions, no drugs or electrodes. From my research, I have learned that fMRI allows scientists to "eavesdrop on the brain at work," a terrifying prospect that makes my hands go clammy and my eyes dart uncontrollably. My mind tries frantically to purge itself of all deranged thoughts, and surely, as a lefty, I must harbor more of them than the usual guinea pig. Following the instructions of a solemn man in a white lab coat, I begin removing all metal objects—earrings, belt, pocket change—all the while sneaking glances at the industrial-sized metal door behind him, certain it must lead straight to the psychiatric wing.

I have come as a representative of the left-handed species, eager to get a peak inside our mysterious cerebral arrangement. I am going to see what a brain looks like when it tackles life "from its right mind," as a famous lefty T-shirt used to say.

* * *

As early as 400 B.C., Hippocrates speculated that the human nervous system is cross-wired, with each of the cerebral hemispheres receiving information from and sending information to the opposite half of the body. This creates a diagonal effect that applies to body movement and touch sensation as well as to hearing and vision— what we see in the right and left sides of our field of view. For example, when fixating on an object, each eye sends the information it sees to the right of the object to the left hemisphere of the brain, while everything to the left of the object is transmitted to the right hemisphere of the brain. The left half of the brain is interested only in visual input from the right side of the world and vice versa. And though the range of vision for each eye overlaps considerably, each eye takes in its own limited view of each "hemifield." [1]

In his treatise *On Wounds in the Head*, Hippocrates wrote, "Most cases have spasm of the parts on one side of the body; if the patient has the lesion on the left side of the head, spasm seizes the right side of the body; if he has the lesion on the right side of the head, spasm seizes the left side of the body." But effects of lesions were different when it came to cognitive functions. An injury near the temple of the left side of the head could leave a person speechless, while a similar injury to the same spot on the right side seemed to leave the patient fully intact.

Our "cognitive functions" include a whole assortment of thinking tasks, from speech and language comprehension to sequential reasoning and emotional processing. Each specialization tends to have its own "specialty shop" in the brain, with counterparts in both hemispheres. But as the brain prides itself on operating efficiently, one shop is usually anointed the "headquarters" for a function. Like a muscle that becomes stronger through greater use, the headquarters is more developed. Speech is perhaps the best example. Unlike vision or movement, speaking does not benefit from symmetry or dueling specialty shops. After all, unlike hands or feet, we have only one mouth, so it helps to have one central command center. [2]

When neuroscientists look at what is common across human brains, often the same regions are tapped for the same speech and language tasks. In the "standard brain," the speech and language comprehension headquarters, which tend to work in tandem, set up shop in the left hemisphere. There are several theories as to why this is the case, many of them linked to the human tendency to prefer the right side of the body, which is also controlled from the left hemisphere. Logically, it follows that in left-handers, this language function should be directed from the right hemisphere, allowing lefties to operate from their "right minds."

This is as much as I knew when I first went to see Dr. Hirsch at the neuroimaging lab. Imagine my surprise when, after reviewing the results from my neuroimaging tests, Dr. Hirsch explained that

I was left-brain dominant. It wasn't enough that I'd been using the wrong hand my whole life; now I was a lefty operating out of my wrong mind.

* * *

Despite what pop psychology has led us to believe, determining a person's brain dominance is no simple feat. Dr. Hirsch arrived at her conclusion only after analyzing my brain scans, a series of snapshot images that were taken as I performed a battery of tests while lying under the enormous white machine. Personality profiles and online surveys tend to operate on the assumption that all right-handers are left-brain dominant, while all lefties are right-brain oriented, a myth born out of the "dual-brain psychology" movement that became popular in the 1970s. In *Drawing from the Right Side of the Brain*, a best-selling handbook first published in 1979, author Betty Edwards wrote that "each of us has two minds, two consciousnesses," with the left hemisphere acting as the logical, straight-arrow rational brain, while the right side is intuitive, creative, and freewheeling. The thinking went that since we're built to be cross-wired, the left-handed had cornered the market on the right brain. [3]

The truth is more complex, of course. We are all designed to use both sides of the brain, and the two hemispheres chatter back and forth constantly. One hemisphere usually dominates—mediating between our intentions and our physical means of expressing them, while converting thoughts into words and actions. Contrary to popular belief, however, the majority of left-handers are considered left-brain dominant, just like righties. [4]

Defining the "Standard Brain"

Before we can understand what makes a left-handed brain different, it helps to begin with a look at how the average right-handed brain is organized, and how science has gone about mapping the geography of the mind. Naturally, there is no one-size-fits-all brain model, but human beings do share a fundamental cerebral architecture and a basic operations chart. The largest part of the brain is the cerebral cortex, the lumpy winding grooves that are visible in most representations of the brain. The cortex itself is actually a thin layer of cells about as thick as a tangerine peel, which sits atop and wraps around the core parts of the brain. Deceptively small with its folded-up coils, the total surface area of the cortex is actually 324 square inches—about the size of a full page of a newspaper. The cortex is divided into the two cerebral hemispheres, which are connected by cables of nerve fibers known as the corpus callosum. Each hemisphere is divided further into four lobes by various sulci (or grooves) and the gyri (bumps) on the surface of the brain.

The left and right frontal lobes of the cortex work together to plan, organize, reason, and solve problems. They also house parts that control speech and movement, including hand control as well as emotional reasoning. The central parietal lobes manage the body's

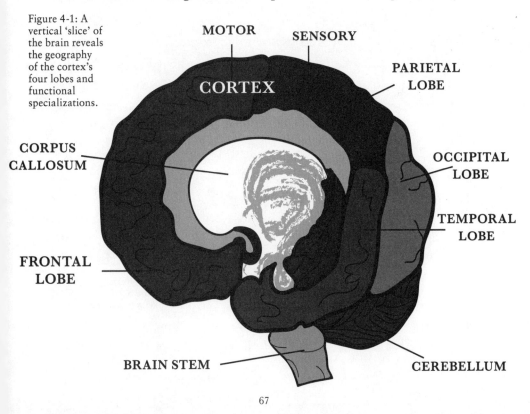

Figure 4-1: A vertical 'slice' of the brain reveals the geography of the cortex's four lobes and functional specializations.

sensations, processing touch, pressure, temperature, and pain. Behind the parietal lobes are the occipital lobes, which analyze the various bits of visual information our eyes bring to the brain, such as color, motion, and form. The temporal lobes, which, as the name implies, lie just under our temples, categorize our perceptions into meaningful groupings (such as "living things" or "musical instruments") and store these perceptions as memories based on their emotional significance. [5]

Neuroscientists tend to focus on the areas that house speech and language comprehension, since these are the tools of the higher-order abilities that facilitate thinking and reasoning. Both speech production and language comprehension (listening and reading) have real estate in the two hemispheres of the brain. The speech centers, located near the temples in the frontal lobes, are small, discrete areas dense with brain cells and grooves. They are called Broca's region, named after the French neurosurgeon, Paul Broca, who first discovered them in 1861.

Not far from the speech headquarters, in a region behind the ears in the temporal lobes, sit two land masses responsible for understanding speech and written language. They're known as Wernicke's region, named after the German neurologist who first identified the area. Together, Broca's and Wernicke's regions comprise the language centers in our brain, and they are usually larger and more developed—or dominant—in the left hemisphere.

Broca was also among the first to suggest that a person's handedness is a mirror of his or her brain organization. He had noticed that when his right-handed patients suffered from brain damage to their left hemispheres, they often lost the ability to speak. The opposite appeared to be true with his left-handed patients. And although he had very few left-handed patients, he concluded that lefties had a "brain in reverse," with dominance in the right hemisphere. This added to the public relations woes of the left-handed, as the right brain had not established much of a reputation back then. For most of the nineteenth century, scientists believed that

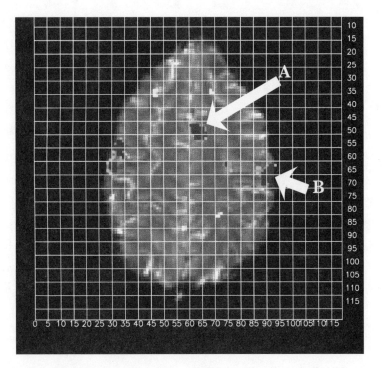

Figure 4-2: A top view of brain activity during a listening exercise, as measured by a functional MRI scan. Right and left are reversed in the final scans so the brain's "executive function," (A) a central organizer for language, activates just to the left of the midline. Wernicke's area, (B) where auditory language is processed, also activates predominantly in the left hemisphere.

the right hemisphere was "the brute brain," performing only low-level functions and taking directions from the "major brain," or the left hemisphere. [6]

Of Two Minds

Brain research would take several big leaps in the twentieth century. In the early 1900s, scientists resorted to carving open the skulls of living patients and poking around until they got a reaction, which helped them map out which region controlled which task. By the 1940s, with the invention of the anesthesia sodium amytal, they learned they could put half the brain to sleep, which turned up new clues as to what each side was capable of on its own. Called the Wada test after the Japanese neurologist who invented it, a sodium amytal injection in the artery that leads to their left hemisphere would leave most patients speechless. Yet two to five percent of right-handers and thirty to forty percent of left-handers became speechless after an injection to their right hemisphere, a jolt to the theory that all righties spoke from their left brains, and all lefties from their right. The differences between left- and right-handers were more complex than previously thought, adding to the mystery that surrounded the trait. What, exactly, could a left-hand preference predict? [7]

It wasn't until the 1960s, when a last-ditch operation for patients with severe epilepsy turned up dramatic findings, that scientists began to revise the major-minor view of the brain halves. After all other measures failed, doctors began to sever the corpus callosum in epileptics, the thick band of nerve cables that connects the two hemispheres of the cerebral cortex. The procedure would stop convulsions from spreading from one side to the other and allow a patient to stay conscious during an attack. After surgery, the seizures would stop, while coordination and daily behavior seemed to remain the same. The patients walked, talked, read, and played sports without any noticeable differences. But the split in their brains prevented the normal exchange of information between the two sides, and in effect, each side was blind to what the other was seeing and thinking. [8]

In 1968, a group of researchers led by Roger Sperry at the California Institute of Technology decided to study these new split-brain patients with a series of cleverly designed tests. What they discovered amounted to a revolution in brain science, and a newfound respect for the left lobe's "silent partner," the right hemisphere. In one early test, a split-brain patient was asked to reach behind a barricade and grasp an object she could not see. When the right hand held the object, the patient described the object correctly: fork. When the left hand reached behind the barricade and grabbed a different object, this time a spoon, the patient fell silent. This changed when images of forks and spoons and ladles

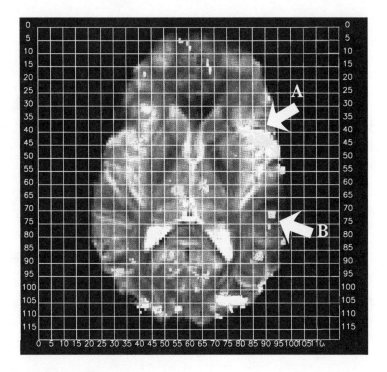

Figure 4-3: During a picture naming excercise Broca's region (A) is dominant in the left hemisphere. Speech processing dominates in the left hemisphere for almost every right-hander and a majority of left-handers, including this one (the author). Since the only auditory processing for this task involves listening to the names as the subject speaks them, Wernicke's area (B) barely activates.

were flashed before the patient. Upon being asked to point to the one she held with her left hand, she correctly identified the spoon. These results were possible because the right hand, which informed the left hemisphere, could communicate what it was holding, but the left hand, linked to the silent right hemisphere, could not summon the word to describe the object. Despite the verbal shortcoming, the brain's visual center still managed to conjure up an image of a spoon; it simply could not transfer the signal from the right hemisphere to the speech translator in the left hemisphere, since the connecting fibers had been severed. [9]

These classic early experiments on split-brain patients laid bare the distinct roles and limitations of each hemisphere acting alone, with each expressing a separate view of the world. It also revealed the full extent of the two hemispheres' interdependence, demonstrating the lengths the left half will go to in order to explain emotional reactions in the absence of the right hemisphere's input. In another experiment, a California woman was asked to look directly at a dot in the center of a screen, and report what she saw. When an image of a nude person was flashed to the left of the dot, processed by only the left eye's visual field, the woman blushed and giggled, but when asked what she saw, the woman replied, "Nothing, just a flash of light," and continued to giggle. When the researchers asked her why she was laughing, her only response was, "Oh, Doctor, you have some machine!" [10]

With language and "higher order" functions controlled by her left hemispheres, the patient could comprehend and articulate only what she saw in her right visual field. Yet it was clear that she also processed what she saw in her left visual field, since she was capable of reacting to the nude photo on an emotional level. Neuroscientists concluded that the verbal side of the brain, in trying to make sense of what's being processed by the nonverbal side, often resorts to fabrications or rationalizations, as it has only a few clues to work with. [11]

The split-brain research also shed light on the hemispheres' different styles of cognition. In one experiment, the patients were asked to match a series of household objects. When shown a cake on a plate, the right visual field/left hemisphere would choose to pair it with a picture of a fork or a spoon. The left visual field/right hemisphere would select images that shared similar shapes, such as a picture of a broad-brimmed hat. This led to the conclusion that the left hemisphere was matching by function, relying on logic and a literal interpretation of the cake, while the right hemisphere focused on the appearance of the object, relying on visual-spatial imagery to find its match. [12]

The most dramatic and sensational of these findings revealed that each of the two sides of the brain seems to house a different kind of consciousness, creating scenarios worthy of Greek mythology. In one famous observation, a split-brain patient swatted at his wife with his left hand while his right hand tried to stop it. Other patients appeared to have two different body images, depending on which visual field (and subsequent hemisphere) processed their reflection in a mirror. When asked to choose what he wanted to be in life from a series of images, one split-brain patient pointed to a race car driver with his left hand, and a draftsman with his right. [13]

* * *

Although the conditions were extreme and the experiments carefully controlled, the split-brain research unveiled the importance of the "brutish" right hemisphere. The Cal Tech group concluded that both hemispheres are involved in higher-level cognition, with each perceiving reality in its own way, and these perceptions are synthesized to form an integrated view through the fibers of the corpus callosum in a normal person. The once-dismissed right brain, they concluded, is actually as complex and powerful as the left brain. The research would earn Roger Sperry a Nobel Prize, and spawn a cottage industry in "right brain" religion—from corporate retreats to advertising tactics to the runaway success of books such as *Drawing from the Right Side of the Brain.* [14]

Inevitably, a backlash followed. Neuroscientists complained that "dichotomania" was oversimplified and oversold. Sperry himself would cringe at the mention of various right-brain gimmicks. And for several years in the late 1980s and early 1990s, most scientists steered clear of the topic. But the 1990s (the so-called Decade of the Brain) brought deeper insights and greater advances in the research, ultimately providing a new respect for the division of functions within the brain—and how they work together. As the research techniques became more sophisticated, conclusions about the two sides became more complicated. In subsequent studies, the hemispheres appeared to borrow from each other's task lists: the right hemisphere was caught in the act of processing simplified verbal constructions, and the left hemisphere was found detecting melodies with different rhythms.

Sign language, a nonverbal form of communication, appeared to be localized in the left hemisphere, much like speech. [15]

As the discussion deepened, so did scientists' hypotheses, one of which suggested that the approach to processing information, rather than the *type* of information, differentiated the two sides. With this model, the left hemisphere was more likely to tackle information presented in a sequence (including sign language), while the right was better designed to grasp many different elements of information as a single whole. The more intricate the interrelations between the elements became, the greater the advantage of the right side. [16]

Incorporating all of these new theories, Russian psychiatrist Vladim Rotenberg and his colleagues proposed that the difference between the hemispheres' two strategies of thinking could be reduced to opposite modes of organizing the "contextual connections" between information. The left hemisphere approach is to organize symbols and signs to create "a strictly ordered and unambiguously understood context." For example, understanding the logic of a statement based on how the information is ordered: "All men are mortal; Socrates was a man; therefore Socrates was mortal." [17]

The approach of the right hemisphere is broader, encompassing a wider net of associations and capturing an infinite number of contextual connections to form an "integral but ambiguous context." In other words, taking in the whole of a stranger's face, including the emotions that it might be expressing, requires an assessment not just of individual features (a triangular nose, for example), but how the features combine to reveal a mood, based on previous experiences with similar faces and expressions. "Individual facets of images interact with each other on many planes simultaneously, much like the connections between images in dreams or in a work of art," explains Rotenberg. He concludes that this strategy of thinking would be advantageous for tackling complex and contradictory information, those things that can't be reduced to a single definitive truth. [18]

The Right Hemisphere's "Reality Check"

In his book *Phantoms of the Brain*, British neuroscientist V. S. Ramachandran recalls visiting a woman in a rehabilitation center in Oxford, England. She had suffered a right hemisphere stroke that had paralyzed the left side of her body. At one point the doctor lifted the woman's paralyzed left arm, held it before her face, and asked her whose arm he was holding. "What's that arm doing in my bed?" the woman asked, startled. Ramachandran asked the patient to explain whose arm it was. "That's my brother's arm," she responded matter-of-factly. Her brother was somewhere in Texas, as were his arms. [19]

The patient suffered from a rare condition called soma-toparaphrenia, the denial of one's own body parts. It is part of a larger disorder known as anosognosia, whereby sane, mentally lucid people become unaware of their illness. In *Phantoms of the Brain*, Ramachandran and author Sandra Blakeslee recount stories of several cases of right-hemisphere damage that result in what seem to be an acute case of denial. "Watching these patients is like observing human nature through a magnifying lens," says Ramachandran. "I'm reminded of all aspects of human folly and of how prone to self-deception we all are. For here, embodied (in one patient), is a comically exaggerated version of all those psychological defense mechanisms that Sigmund and Anna Freud talked about . . ." [20]

While the Freuds offered the psychological explanation for denial, Ramachandran offers a neurological theory. Among the many divisions of labor between the two cerebral hemispheres, he suggests, are two fundamentally different ways of coping with the multitude of sensory inputs the brain must process. "The left hemisphere's job is to create a belief system or model and to fold new experiences into that belief system. . . . The right hemisphere's job is to play 'Devil's Advocate,' to question the status quo and look for global inconsistencies." At a certain point, when the new information clashes too dramatically with the old, the right hemisphere serves as updater. If the updater is not working, a person is left clinging to the old information, which (often conveniently) allows him or her to remain in a state of denial. [21]

Cognitive psychologist Stephen Christman and his colleagues took a different look at the right hemisphere's role as "reality checker" in a series of 1997 experiments at the University of Washington. They showed subjects a series of body silhouettes that ranged from extremely thin to extremely fat, and asked them to indicate which silhouette most closely matched their own body shape and size. Then, calculating each subject's actual body mass index, they looked at the discrepancy between perceived and actual body size. The discrepancy was nearly twice as big in those who were "strongly handed," a group Christman had previously found to be limited in cross-hemispheric interaction. The strongly right-handed, in particular, with less access to their right hemisphere's perceptual capabilities, appeared to be susceptible to a more distorted self-image. These results might help to explain the split-brain patient who, when asked what he wanted to be someday, pointed to a race car driver with his left hand and a draftsman with his right. Perhaps his left hand (and left hemisphere) lost access to his reality updater in the right hemisphere. [22]

This responsibility for realism may explain the right hemisphere's role as the control center for the "withdrawal" response and the processing of negative emotions. One of the more news-making discoveries that came to the forefront of the discussion during the 1990s was that the brain's right and left hemispheres tend to divide responsibilities for negative and positive emotions. Imaging studies from the Laboratory of Affective Neuroscience at the University of Wisconsin revealed that the brain's left prefrontal cortex was more active when positive memories involving happiness and amusement were invoked. By contrast, the right prefrontal cortex lit up when negative or inhibiting feelings were produced, such as viewing images of an angry face. [23]

Facial expressions are also influenced by the hemisphere's divided emotions, as fine motor control of the left and right side of the facial muscles is controlled by the opposite hemisphere. (If you're a lefty, chances are you sneer better using your left upper lip. And the next time you see lefty actress Sarah Jessica Parker on-screen, watch what she can do with her left brow.) A new imaging method that studies facial movement in three dimensions discovered uneven movements of the face during expressions of happiness and sadness. Researchers observed more left-sided facial movements during all emotional expressions, but particularly during expressions of sadness. [24]

The Cerebral Organization Chart

In most of us, both hemispheres of the brain contribute to everything we do, albeit in different ways. While we're busy thinking and reasoning, different regions in both halves of the cortex become activated, a process captured by neuroimaging scans that reveal blood flowing to each region in surges. In the average brain, this activation process begins at a central motor command center located in the left cerebral hemisphere. Like two separate companies that have merged, the brain's "headquarters" are usually consolidated here, helping to streamline orders and prevent decision-making gridlock.

Within this headquarters, there are two levels of responsibility: the higher- and the lower-order functions. The higher-order "executive suite" is responsible for the Big Thinking tasks such as formulating goals and thoughts, converting those thoughts into language elements, planning and directing physical actions, and focusing our attention toward one hand at a time. The nearby lower-order functions carry out the commands from the higher-ups, including translating words into sounds (speech) and converting plans into physical actions (movement). These "final outflow" functions often operate at a level of automation, which means they can take care of themselves without

Figure 4-4: A "vase-face" image that neuroscientists use to study visual perception. Subjects that rely more on their left hemisphere usually see the white vase. When the right hemisphere predominates, people are more likely to see the profiles of two faces.

involving the neurological higher-ups. This is why we don't have to consciously decide to breathe, blink, comprehend the words we hear, or contemplate which leg to move while walking. [25]

The higher- and lower-order functions differ in terms of how much of the brain's neural landscape they engage. Higher-order language tasks such as selecting words and sequencing them to form sentences are usually directed by one hemisphere but tap different regions in both halves of the cortex. Yet the lower-order task of converting the selected words into sounds requires a concentrated and highly specialized speech center in one discrete area of the brain—usually in the left lobe, adjacent to the motor control center. "Generating sentences requires that we activate much more than the areas directly involved with the mechanics of speech production," explains Michael Peters, Ph.D., professor of psychology at the University of Guelph in Ontario and an editor of the journal *Laterality*. The tone in our voice, the emotional expression, the volume and modulation, the accompanying facial expression and gestures, a joke or a pun thrown in—all of these require input from many different regions of the brain, in the left hemisphere and the right. "And all the time, while we are talking, we assess whether the listener pays attention and whether we have to modify what we say and how we say it on the basis of their reaction," explains Peters.

There is evidence of this process in my own brain scan experiment. When I was asked to think of a story, broad areas in both hemispheres became active, which is expected. But when it came to articulating these thoughts, the activity became more concentrated, and the scan revealed a dark, dense nexus of activity in the left hemisphere's speech center, the infamous Broca's region. Like all fine-motor instructions, speech has to be precisely formulated and sequenced. The speech apparatus—the two sides of our jaw, mouth, throat, and larynx—requires tightly controlled orchestration, so it helps to have the command center for generating words adjacent to the command center that executes the production of their sounds.

Executive Language

Before we can speak and write, we must learn the meaning of words, formulate thoughts, and translate those thoughts back into words and sentences, a process known as higher-order language. In most people—meaning almost every right-hander and sixty to seventy percent of left-handers—this process is dominant in the left hemisphere of the brain, where language is received and interpreted, and where words are then selected and sequenced. Ultimately, higher language is not solely based in the left hemisphere, since it is a relatively diffuse operation that taps into many different supporting functions, or "subprocessors," such as visual imagery or intonation, a few of which typically reside in the right hemisphere. [26]

A good example of this can be seen in patients who have suffered a right-hemisphere lesion. They are likely to lose the ability to grasp some of the more elusive aspects of language, such as abstraction,

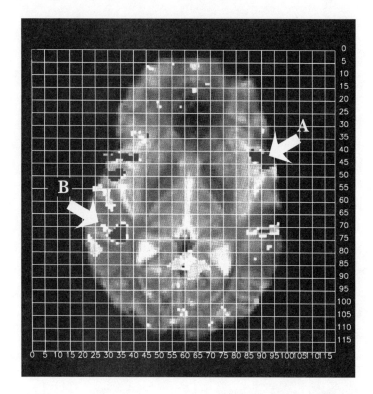

Figure 4-5: During a synonym generation task the right hemishere's Wernicke's region (B) dominates in this left-hander. Basic speech is still dominant in the left-hemisphere as evidenced by Broca's activity (A).

context, and metaphor, as well as the nuances of semantics and tone, which often require interplay between the two hemispheres. This condition causes patients to relate to language on a purely literal level, so that when scientists present a sentence such as "the man had a heavy heart" to them, they are likely to point to a picture of a man lugging an enormous heart up a hill. In contrast, a person who has suffered left hemisphere damage and must rely solely on his or her right hemisphere for interpretation is more likely to point to a picture of a man who is obviously sad. [27]

Damage to the right hemisphere also interferes with the ability to understand sarcasm or the context in which something is said. If a patient with a right-brain lesion hears someone say "well done" in a sarcastic tone after they've flubbed a task or an exercise, they may

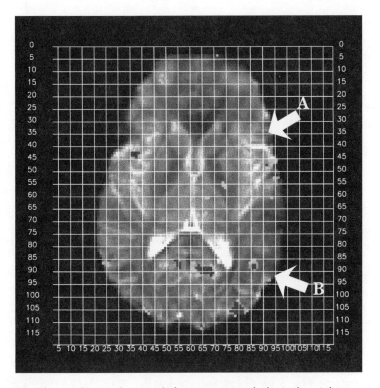

Figure 4-6: When a subject "rehearses" before an exercise, the brain doesn't have to work as hard, as evidenced by the faint activation patterns during an internal (silent) storytelling exercise. Note that Broca's (A) and Wernicke's (B) still appear to dominate in the left hemisphere—despite the subject's left-handedness.

become confused and miss the comment's intended meaning. While the left hemisphere can process the literal meaning of language on its own, it needs the right half to comprehend the finer elements of heard speech—intonation, voice contrasts, and the subtleties of sounds (the difference between "gu" and "ku" for example). People who've suffered right-hemisphere brain damage tend to speak flatly, with little emotion or tone. In some studies, they also have trouble reading faces, which has been attributed to a larger problem with interpreting people's emotions or contextual cues, according to Robert Ornstein, a neuroscientist and author of *The Right Brain*. Ornstein speculates that right-hemisphere damage could block the interpretation of any complex visual display, including the minute changes in facial coordinates that occur when a person is speaking or emoting. So if someone smirks slightly or cocks an eyebrow after delivering a deadpan sentence, a person with limited right-hemisphere function might not pick up on the cues, and will interpret the sentence literally. [28]

The Speech Output Center

We have only one mouth and two hands, and they perform very different functions, but handedness and speaking are closely related. For one thing, they share some common geography in the brain, as their headquarters tend to be adjacent to each other in the same hemisphere—at least in the average brain, the "standard model" shared by nearly every right-hander and a majority of left-handers.

A "movement" action of the mouth requires coordination of both sides of the speech machinery—the lips, jaws, tongue, and larynx. Yet this complex apparatus is ultimately directed by the speech headquarters, located in the left hemisphere in the standard brain, which sends its signals through the cranial nerves on both sides of the brain. The flow chart for speech is fairly elaborate. To speak and read a written word aloud, information must first reach the primary visual cortex in the back of the brain. From there, the letters and corresponding sounds are transmitted to the speech-reception region (the seat of Wernicke's area), from which information travels to Broca's area—the speech production regions—and on to the primary motor cortex. The same basic route is followed when a spoken word is repeated aloud, except instead of starting in the visual cortex, the process begins with the auditory cortex, which then relays the information to Broca's region, and finally to the primary motor cortex—which also controls hand movements. [29]

The Motor Directors

Like language and speech, the movements and skills of the hands are governed by a similar higher/lower division of labor. Fine-motor skills, right down to the slightest muscle extension, require the input of several cerebral departments, beginning with a "general motor planning" department, which issues commands ("must pick up grocery bags") to both sides of the body. For most people this function is controlled by the left hemisphere, working from an area near the speech headquarters.

The right hemisphere still plays a role in the early stages of all motions, and there's evidence to suggest that it plays an even larger role in the left-handed. As with higher-order language, the motor-planning headquarters relies on regions of the right cortex to serve a support function, such as helping to execute movements. We know this because of a motor disorder known as apraxia, which literally means "unable to act or carry out." Patients with apraxia can move their muscles perfectly well, but they have trouble planning and sequencing specific movements, executing a throw or getting a spoon into their mouths. In right-handers, apraxia is strongly linked to left-hemisphere damage, but subtle problems in planning and executing movements can also be seen after damage to the right hemisphere. In left-handers, brain lesions in the right hemisphere cause greater problems in carrying out sequences of movement than is the case for right-handers with lesions in the right hemisphere. [30]

Even two-handed activities such as peeling an apple or playing the violin are directed by the one dominant motor-planning headquarters, and in almost every righty and a majority of lefties, it resides in the left hemisphere—playing maestro to both hands, telling the dominant hand what to do and the other hand when to do it. When a lefty peels an apple, the motor-planning department calls attention to the left hand because it's the hand most adept at manipulating the peeler. "As a broad principle, we don't pay as much focused attention to the object we are acting on (rotating the apple while we are peeling) as to the hand that does the acting (the hand that wields the peeler)," explains Professor Peters. "Both hands are active in a complementary way and which hand we choose for which task says much about our handedness."

The actual execution or "outflow" of our movements—the wrist and finger manipulations required to flick a peeler and turn an apple—activates more specialized regions of the brain, and this is where the cross-wiring effect comes into play. Each hemisphere has both a gross- and fine-motor output department, and these motor outputs are responsible for executing the arm and hand (and leg and foot) movements of the opposite side of the body. The gross movements of our limbs, such as lifting, throwing, and kicking, tend to be controlled by both motor-output departments, which makes sense given that these larger movements often cut across both sides of our body. The fine-motor movements, especially the grasping movements of the fingers for tasks like handwriting, are more likely to be controlled by the motor-output center in the opposite hemisphere. In left-handers, the dominant fine-motor center resides in the right hemisphere. [31]

The nondominant hand also has a fairly active motor-output center, since many activities require the use of both hands extensively, and our lesser hand needs to be qualified enough to carry out a wide range of tasks—on its own or in support of the preferred hand. This support function can be just as complex and demanding as the dominant hand's activity (think of the finger movements on a string instrument), and so the area of the brain that drives it must be just as specialized.

Should the dominant hand be lost for any reason, the other is capable of learning tasks once handled by its counterpart, usually through physical therapy, regular use, and ongoing practice. When a stroke knocks out the systems that provide the final motor-outflow control for one hand, the systems for the other hand are poised to take over.

Directing "Gross" Motor

Microsoft founder Bill Gates is one of those peculiar lefties who writes left-handed but throws a ball with his right hand. We know that in Western countries, there are actually a large number of left-handers, up to 30 percent, who divide the labor, writing with their left and throwing with their right. Michael Peters, the first neuroscientist to study this subgroup, sees them as a critical piece in the handedness puzzle, shedding light on the differences between fine and gross motor movements. "Here we have a group of individuals who pick and choose what hand they use for what activity, and they seem to prefer the right hand for whole arm, power activities while using the left hand for fine finger coordination," explains Peters. This division of labor also exists in lesser species such as lobsters, which use their larger "crusher claw" to conquer and hold their food while relying on their slimmer, more delicate claw to extract meat from the prey and to bring it to their mouths. [32]

The separation of gross- and fine-motor functions can be useful in humans as well. "Think of a parent holding his or her infant in one arm while feeding the child with the free hand," offers Harris. In contrast, consistent left- and right-handers—those who write and throw with the same hand—put all their eggs in one basket, vesting both strength and finesse activities in the preferred hand, rather than distributing them across the two hands. [33]

"There is a temptation to confuse the way in which muscles are controlled with the precision of the movement," says Peters. "When we use our fingers, for example, we are using the most highly developed part of the motor system, the part that allows for rapid, precise, skilled movements. This system is anatomically different from the one that is used to control the large muscles that are involved in maintaining posture, because for fine manual movements we use motor tracts that extend directly from the motor cortex to the motor cells in the spinal cord, while for large postural muscles more indirect tracts are used that synapse several times on the way from the brain to the spinal cord." [34]

Peters points out that despite the anatomical differences between fine- and gross-motor control, both types of movement rely on the same brain networks to focus attention on the goal of the movement, monitor performance, and learn from errors. "For instance, throwing requires two different types of movement: the strength component that involves the arm, the shoulder, and the entire body, and the finesse element that requires the delicate and precise movements of the hand as it directs and releases the ball. If these two elements are not perfectly coordinated, the outcome suffers. So functionally speaking, 'fine' and 'gross' motor movements answer to the same master. They are both directed by the same networks that guide movement in producing a cascade of coordinated muscle contractions that begin with the musculature of the torso and limbs and end with the movements of the fingers." [35]

The Twenty-First Century View: Brain as Orchestra

As recently as the mid-1990s, many neuroscientists still viewed the brain as a neat and orderly yin-and-yang symbol. Studies of cognition reinforced the idea that the left hemisphere worked more like a zoom lens, homing in on fragments and fine details and tackling mental tasks that require a series of steps, such as ordering words to communicate a thought, or linking that cake to the fork needed to eat it. The right hemisphere was the wide-angle lens, pulling back to capture the big picture, in full panoramic view, the better to determine the relative position of objects (a cake looks a bit like a hat), or words within their larger, metaphoric meaning.

This relatively simplistic view of the brain came apart with the introduction of the fMRI, which opened up exploration into the different roles of the two hemispheres, revealing the full extent of how they work together to create an ongoing symphony of movement, thoughts, emotions, and behavior. With the fMRI, patterns of nerve-cell activity are observed through changes in blood flow—the more active the neurons, the more oxygen and glucose has to be delivered to them through the blood supply. In the first fMRI experiments on normal subjects—those without lesions or disconnected hemispheres—patterns of nerve-cell activity revealed that both sides of the brain played an active role in each cognitive task. When the average right-handed subject was asked to listen to someone speak, for instance, the left hemisphere would surge with activity in the areas that decode speech sounds. The right hemisphere would show comparatively less blood flow than the left in the counterpart regions that deal directly with speech sounds, but greater blood flow in the areas that respond to the emotional tone of a speaker's voice. The "live action" images made it clear just how engaged both hemispheres were in listening to speech, with each contributing its own specialization, shaping how we receive what is communicated to us. [36]

Some of the same underlying patterns were found: the left hemisphere seemed to tackle the core aspects of speech, such as grammar and word production, while the processing of emotional intonation and metaphors tended to send the oxygenated blood to the right hemisphere. Interestingly, the brain scans revealed that the verbal left hemisphere could also play a role in spatial tasks, albeit a different one. Certain regions in the left lobe surged when a person grappled with specific shapes or objects at specific locations, while the right hemisphere activated when they were conceptualizing a more general sense of space and distance, or when they were asked to group shapes together. [37]

Today, most neuroscientists view the brain as a coordinated, integrated system of networks, with almost every mental task shared across the brain and each side complementing the other. The specialties of each hemisphere are viewed as relative as opposed to absolute; one may be more quickly and more intensely engaged for a semantic task than the other, firing up a larger or denser blot in the brain-scan images, but the correlate in the opposite hemisphere still showed some activity. Instead of a single hemisphere managing all aspects of a single cognitive task, discrete subprocessors or "modules" within each hemisphere are dedicated to different parts of the individual task, with the left and right hemisphere processing simultaneously. [38]

While rejecting the notion of an overall processing dichotomy between the two hemispheres, neuroscientists today agree that the two hemispheres do have contrasting styles of working and that trends and patterns can be found to distinguish the two. When a subject listens to a sentence, for example, a brain scan will show surges of activity in the auditory regions of both hemispheres, but one side—usually the left hemisphere—will show activity spread out across a larger area, indicating that several nearby "satellites" were tapped. This is most likely because the left brain's subprocessors are taking charge of syntax and grammar, sequencing the words in the sentence and decoding their precise literal meaning. Scientists postulate that the smaller, less intense combustion of activity in the right hemisphere is the "wider context" subprocessor, which flips through a broad-ranging internal thesaurus to generate multiple meanings of each word, matching metaphors and decoding tones such as sarcasm and fear in the speaker's voice. [39]

An interesting parallel can be found in visual processing. There is a familiar image of a shapely white vase that sits on a black background, and when most people are asked to describe the image, they will describe the vase. But if the right hemisphere becomes engaged more readily and more intensely, the subject will see the image differently: the outline of the vase, the black "backdrop," is actually two profiles of faces looking at each other. The left hemisphere will light up to process the vase figure—or foreground—while the right hemisphere kicks in to decode the background. [40]

Although the bulk of evidence still suggests that the left hemisphere is designed to focus on local processing (the vase or foreground) and the right hemisphere is more inclined toward the global (the whole picture, including the background), the way in which these processing differences play out depends on the nature of the task. For example, when it comes to recalling the details of a trip, the left hemisphere will see activity when the person remembers the sequence of events, but the right hemisphere will surge when he or she reminisces about where and when each event occurred, and the visual details that were present. [41]

The Lefty Brain

When I returned to Dr. Joy Hirsch's neuroimaging lab, the results from my brain scan experiments were laid out on a table, a gallery of brightly blotted images resembling an elaborate relief map. Gradations of colors somehow bled from the white elephant of a machine to the computers in the control room, which spit them out as calibrated diagrams of what look to be a sloppy symphony. Big blobs of red intensified to become smaller concentric circles of orange that would close in on bright yellow peaks—all located at different points depending on the exercise.

Technically speaking, the fMRI does not directly measure blood flow, but rather the change in "magnetic susceptibility" associated with increased blood flow to an activated region. When the brain is working extra hard, a lot of blood is flowing, and the signal becomes stronger, translating to a brighter color on a brain map.

Looking at the images from my first task, Dr. Hirsch pointed to a small, dense yellow patch of activity in my left hemisphere—Broca's area—which activated when I had to name objects displayed on pictures—chair, umbrella, cat. In the next exercise, when I listened quietly to instructions, a similar blot appeared not too far away on the same side—Wernicke's area. "When push comes to shove and you're using only basic elements of your language system, you're fairly biased towards your left hemisphere."

Just as I was beginning to accept my new left-brained identity, Dr. Hirsch pulled up the results from my synonym generation exercise. "We got some surprising findings here," she said, pointing to the orange and yellow pond across the divide in the right hemisphere. The synonym task is expected to yield the same basic activation patterns as picture naming and passive listening, only with more intensity, and with greater use of the higher-order language executive coordinators, given that it takes more work. But this time, the surging oranges and yellows took a different course. A big patch of yellow appeared in my right hemisphere's Wernicke's area. (Recall that Wernicke's is where we conceptualize language and process its meaning.) "You're much more bilateral here," Dr. Hirsch explained. "You're still using your left hemisphere, but you call upon the right hemisphere more readily here, unlike in the other tasks."

My patterns seem to parallel those of another writer who volunteered to have his brain mapped by Dr. Hirsch. In 1999, her lab took on right-hander Stephen S. Hall for a piece he was writing for the *New York Times Magazine*. As is expected in a righty, his brain demonstrated clear-cut left hemisphere dominance for speech and language. But when he was asked to generate synonyms, what he referred to as "the great thesaurus expedition," the right side of his brain "lighted up like a neon sign on a cheap diner," as he described it. Parts of his visual cortex, Broca's speech production center, as well as an executive organizer and coordinator (known as the medial frontal gyrus) all surged with activity in the right hemisphere. Dr. Hirsch speculated that Hall had a natural predisposition to use the right side of his brain, while also giving it frequent exercise throughout his writing career. (Contacted in 2004 for this book, Hall also explained that he has a left-handed son, evidence that he may share predispositions toward a "lefty brain," as we shall see in the next chapter.) [42]

The fourth and final exercise, internal storytelling, was invented by Dr. Hirsch to get a better sense of just how the language system orchestrates its different components. My results were not so good. "Did you cheat here?" Dr. Hirsch asked, pointing to the relatively weak and spotty display of small red patches. I panicked for a moment. "Was there a right story to tell?" I laughed nervously. "We suspect you prepared in advance," she said. Oh, that. I did know ahead of time that she was going to ask me to recall the details from my previous day, and it was not a good one—that is to say, nothing I wanted the man in the white lab coat to know about. Instead I decided to pick the previous day and pulled up the events while I was waiting for the test to begin. "Rehearsing" is a neuroimaging no-no. The brain doesn't have to work so hard once it has rehearsed, and the activity tends to be weaker and therefore harder to read. But Dr. Hirsch was able to detect some faint patterns from my underactive brain. "There is a nice signal in the intention side," she explained (perhaps because I was trying not to recall the bad day), and it appeared that my "executive" or higher-order language system activated once again on the left side. "Yet the left hemisphere is quick to call on the right hemisphere," Dr. Hirsch pointed out. "We expect internal speech, or storytelling, to have a broad activation. In your case, you specialize on the left side, but you're calling upon other supporting areas in the right side, more or less depending upon the task. That variation is huge within you."

Examining all of the results, Dr. Hirsch concluded that I was both left-hemisphere dominant for basic language and bilaterally dominant for more complex language functions. This sounded somehow reassuring, if a bit inconclusive. Does it explain why I can't make up my mind? Will I always be at war with myself? Not exactly the domain of the fMRI machine. I couldn't even get an answer on whether I was a "normal" lefty.

Because neuroscience labs are primarily in the business of looking for patterns and mapping individual brains, they do not tend to focus on variations. Many of the contradictions that turn up in the research do so because the specifics of a task can inspire different responses, making it difficult to compare findings. For a synonym exercise, for example, nouns and verbs can churn out different results. "Research is usually designed to find commonalities," explains Dr. Hirsch. "We like absolutes, black and whites. But we greatly underestimate the amount of variations."

These "variations"—from absolutes and from within and between individuals—turn up throughout the literature on brain research. Yet brain-imaging studies have uncovered interesting differences between left- and right-handers, from the density and firing of neurons to the supply of oxygenated blood to different regions— an indicator of which functions are tapped for different tasks. It is up to the scientists who cross-reference brain studies with handedness to offer other theories as to why an internal thesaurus would break off from the dominant language system in the left hemisphere and end up in the right. Some of these scientists have suggested that the brains of the left-handed—and the right-handed who carry similar genes— display any number of cerebral combinations, with some functional specializations (writing, reading, even emotional processing) laying roots in the left lobe, and others in the right lobe—a grab bag of brain landscapes. [43]

"Unlike the strict asymmetry witnessed in right-handers' brains, the left-handers have lost the bias," explains Daniel Geschwind, a geneticist and neuroscientist at UCLA. "They may still have strong left hemisphere dominance, but the wiring is less strict or confined, and this opens up their cerebral development to a range of outcomes." [44]

This idea of left-handers exhibiting "random cerebral variation" was put forth by Chris McManus, a professor of psychology and medical education at University College London and author of *Right Hand, Left Hand*. His theory suggests that, in left-handers—and right-handers who are predisposed to the same neurological tendencies— the brain's "submodules" can develop in unexpected places. This allows components of language, including writing and reading, to have "minds" of their own and develop as separate entities, resulting in separate hemispheres. For example, visuo-spatial reasoning and emotion, right-hemisphere modules in the standard brain, could end up in the left hemisphere of a left-hander. [45]

One case that reinforces McManus's theory of rearranged modules is that of "VJ," a University of California San Francisco research subject whose brain hemispheres were separated in 1995 after years of severe epileptic seizures. After the surgery, the left-hander underwent the usual series of split-brain experiments, but her responses shocked neuroscientists and made headlines. When words were flashed to VJ's left hemisphere, she could read them, speak them, and spell them without any hesitancy, but she couldn't write them down. When the same words were flashed to her right hemisphere, she would be stumped, unable to read or speak what she saw. She told researchers, "I think there's something there, but I can't tell what it is." Then, to their shock, after claiming she could not decipher the word, she would write it down—using her less-skilled right hand. The findings challenged the notion that the elements of language—writing, reading, and speaking—represent a monolithic enterprise, developing in sync under one roof in the left hemisphere. [46]

The Case of the Closet Lefty

Often graduate students sign up to serve as a control for various brain experiments. To qualify as control-group material, however, a subject has to be consistently right-handed. Neuroscientists count on righties to provide them with more predictable brain organizations.

Then came Michelle. When Dr. Hirsch and her colleagues at the fMRI lab pulled up the images from the grad student's functional brain scans, they were completely stumped. During the language exercises, when Michelle was asked to speak, listen, and generate synonyms, her brain surged with activity, but not as expected. "Her right hemisphere was screaming," Dr. Hirsch recalls. "We were all thinking, 'What is going on here, you're a right-handed person!' We sat her down and asked her, are you really, really right-handed? Were you ever switched as a kid? And she was sure of it. She had always been right-handed. One of the lab scientists said: 'I don't believe you.

Call your mother.' And so Michelle called her mother and asked if she was born left-handed."

After some prodding, the mother confessed. Michelle had been born in Korea, a country that even today frowns upon left-handedness, especially in girls. From an early age, Michelle started using her left hand, but her mother actively discouraged it. Michelle had no memory of this, and no inkling that she was ever left-handed. [47]

Between six and ten percent of adult Americans considered right-handed based on handwriting were switched as young kids—and this includes only those who are aware that they were converted. How many other so-called righties are like Michelle? More importantly, how does the brain organization of these closet lefties compare with unconverted lefties and righties? [48]

In 2002, German and British neuroscientists studied the brains of "switched lefties" to see how their brains were lateralized. They rounded up eleven "innately left-handed" adults who had been successfully converted to right-handed writing, as well as eleven who had been consistently right-handed to serve as the control group. Using positron emission tomography, which measures changes in cerebral blood flow using electrodes, they monitored the subjects as they wrote common words using their right hands. In the innate right-handers, handwriting led to activation of the left hemisphere's motor and sensory integration input areas, which was expected. Yet the converted left-handers demonstrated more bilateral activation patterns—with distinct "hot spots" in the right hemisphere's motor, sensory-somatic, and audio-visual perception areas. These findings offered evidence that even after decades of right-hand writing, the left-hander's brain continues to be wired differently. [49]

During further discussion, the scientists speculated that the right-hemispheric activation in converted left-handers could reflect the suppression of unwanted left-hand movements, and that it may indicate that the area of the brain responsible for hand movements is more hard-wired than previously thought. No amount of exercise given to the nondominant hand could convert the brain to become truly "right-handed." [50]

If left-handers can be left-brain dominant, and right-handers can be right-brain dominant, then what do our hand skills really tell us about our brains? Is there anything predictable about a left-hander's head? Surely there must be some neurological traits that separate the aberrant lefties from the conforming righties?

Unpredictability may be the one predictable trait all lefties share. Brain studies across the continents have lent credence to the idea that left-handers as a group display more varied cerebral organization, with a greater range of activation patterns or a more diffuse spread of neural activity. Even among "standard brain" lefties, those whose brains appear to resemble right-handed textbook brains in every way except motor dominance, there is more variation from the expected pattern of left-hemisphere dominance. [51]

While variations are expected in all brains, bigger variations are expected in the left-hander's brain, a fact that has been reinforced by studies that examine it from inside and out. Dan Geschwind led a team of researchers in a computer-aided study that measured the brain scans of seventy-two pairs of identical male twins, along with sixty-seven sets of fraternal twins serving as a control group. As expected, they discovered that the brains of identical right-handed twins were very similar in size and structure. The left-brain hemisphere was larger in volume size, an indicator of language dominance—as Broca's and Wernicke's can expand to three times the size on the dominant side of the brain. [52]

The results were different when one member of the twin set was left-handed. In these instances, the brains of both members were more symmetric or balanced: the left and right hemispheres were more equal in volume, an indication that language functions were less restricted to the left hemisphere and more likely to "take flight" and end up across the great divide between the hemispheres. The twins inherited the same neurological organization but ended up with a different handedness outcome. How this comes to pass is a subject for the next chapter, but the telling variable here is the organization of the brain. The left-handed, and those who share their genes, have more flexible brains, explains Dr. Geschwind. "Their language functions are more likely to be parceled out, distributed across the two hemispheres. This may put them at risk for some neurodevelopmental disorders, such as dyslexia, but it also provides some advantages." We'll examine these advantages in greater depth in coming chapters, but from a neurologist's perspective, one of those advantages is greater "plasticity," which means the brain is quicker or better at recovering from damage—be it an injury or a stroke. Studies of war veterans recovering from brain injuries found that the left-handers were more likely to regain speech and movement quicker than were their right-handed counterparts with similar injuries. [53]

A more developed "nondominant" motor headquarters may have an evolutionary purpose. In the event that one-half of the brain—or body—becomes injured, the other half is more poised to take over. Matthew David Scott, a paramedic who became the first successful hand transplant patient in 1999, lost his dominant left hand in a fireworks accident in the 1980s. He quickly learned how to write with his right hand, and adapted remarkably well. While his new transplanted left hand has made great strides—it threw out the first pitch at a Philadelphia Phillies game three months after it was attached—Scott continues to write legibly with his right. [54]

The Left Hand's Headquarters

Regardless of their language variations, the left-handed all share one fundamental neurological underpinning: their hand-control center, or primary motor cortex, is bigger, denser, and more sensitive in their right hemisphere. The circuitry is more intricate, and the pathways more paved, allowing for increased blood flow to this destination, according to Michael Peters, who has collected and analyzed brain scan results of left- versus right-handers from neuroscience labs throughout the world. [55]

Building on these findings, Katrin Amunts and her colleagues at the University of Düsseldorf, Germany, also collected brain scans from left-handers and right-handers whose handedness had been carefully measured. They looked at the primary motor cortex, the part of the brain most directly involved with the control of movement. If there were clear differences between the brains of left- and right-handers, this is where they would be found. They zoomed in on the large channel separating the frontal and parietal lobes, what is known as the central sulcus. The deeper the sulcus, the more developed the motor cortex is. This is not because it contains more brain cells, but because it is richer in synapses and axons, the meatier connector cells that take up more space. "When the brain first develops, the areas where there are the most vigorous proliferation of cells and their connections will see the most dramatic folding of the cortex, and therefore deeper sulci," explains Peters. In other words, the deeper the grooves, the more widely spaced and heavily interconnected the neurons—an indication of the complexity of interactions between nerve cells. [56]

Given that the motor cortex in the left half of the brain controls most of the movement in the right body, we'd expect the left-central sulcus of right-handers to be deeper than its counterpart. Sure enough, this was what the researchers found. They also found the converse for their left-handers: the right-central sulcus was deeper than the left. However, in keeping with a pattern that is often observed when comparing the brains of left- and right- handers, the differences between the two sulcus depths were not as pronounced in the left-handers, suggesting that the differences between the motor cortices in left-handers are less pronounced. "The meaning of this is currently unclear," explains Peters. "One can speculate that the region controlling the non-preferred hand is more developed in left-handers than in right-handers, with the region controlling the preferred hand being comparable for the two groups."

As if all of this is not complicated enough, there's still the chicken-or-egg question. Do right-handers have a larger motor area in their left hemisphere because they use the corresponding hand more, or do they use their right hand because the left hemisphere's motor-hand area is larger? Studies of musicians have found that hand-motor areas in the brain can become larger as a result of intense use of one or the other hand. This could explain why the left-handed have more equally developed motor cortices in both hemispheres. They have no choice but to give their nondominant hand more exercise, adapting to right-handed appliances, tools, sports equipment, and computer mice. [57]

The Role of Hemispheric "Cross Talk"

Greater activity and development in the opposite hemisphere may also explain why a left-handed person is more likely to have a larger corpus callosum, the bundle of nerve fibers that connects the two hemispheres. Research on cadavers in the early 1980s found that the left- and mixed-handed tended to have thicker bridges between the hemispheres. More recent behavioral studies have also concluded that the left-handed have more connections between the two halves, an indicator of greater hemispheric "cross talk." This notion of increased interhemispheric communication has been examined by cognitive studies that present stimuli separately to the left and right hemispheres—words, colors, or images, for example—and monitor the extent to which the stimuli are processed together or separately. The results from these experiments have led neuroscientists to speculate that people who are mixed-handed (a lefty who throws right-handed, for example) demonstrate the greatest communication between the two hemispheres. Mild righties who have some left-handed proclivities would also fall into this category. But given that all lefties are forced to use their right hands on a regular basis, they are more likely to call on both sides, so odds are greater they have more interaction between their two hemispheres. [58]

Christman and his colleagues at the University of Toledo have conducted several studies that use indirect measures to determine the relationship between hemispheric interaction and handedness. In one study, subjects were asked to keep journals recording memorable or unusual events; weeks later, without notice, they were asked to recall the details of the events—without the aid of their journals. The subjects with mixed-handed tendencies demonstrated significantly better "episodic memories" in that they were more capable of recalling the time and place of an event as well as the event's sequential details. Lending support to these findings are brain-imaging studies that reveal that the encoding of episodic memories occurs in the left hemisphere, while the retrieval is more likely to occur in the right hemisphere. So a stronger recall of events—the sequence and context—provides evidence of greater integration in the brain. [59]

To facilitate this communication between the hemispheres, it helps if both hands are working together. Playing the saxophone or violin, for example, requires that the two hands collaborate, and forces them to coordinate their roles through their respective motor-output centers in the opposite hemispheres. A 1993 study of professional musicians found that those who played two-handed wind and string instruments were more likely to be mixed-handed. Piano players, whose hands must act independently, were far more likely to be strongly handed, to the right or left, than the wind or string instrument players. Despite giving both hands plenty of exercise on the keyboard, in everyday activities, piano players tended to use only one hand across all tasks. This "separatism" allows their hands to maintain separate "minds" and limits the interplay between the hemispheres. It also helps to explain why the left-handed, with their heightened levels of hemispheric communication, show up in higher numbers among string and wind musicians than among pianists. [60]

The success of left-handers at strings can be seen in two of the world's most legendary musicians, both of whom played the guitar left-handed, despite the fact that the instrument was not designed for them. Paul McCartney adapted by turning his guitars upside down. Approaching the same problem differently, Jimi Hendrix, who was declared the greatest guitarist in rock history by *Rolling Stone* magazine in 2003, simply restrung right-handed guitars in reverse. [61]

Of course, a surplus of hemispheric interaction is not necessarily all good, since one of the jobs of the corpus callosum is to filter out cross-talk to prevent stimuli overload. Lefties have been known to struggle with tasks that require independent processing by the hemispheres, the simultaneous "pat the head, rub the tummy" drills, where it helps if the left hand really doesn't know what the right hand is doing. In another revealing study, the word *blue* was written in green ink and flashed to different subjects, and the lefties struggled more when it came to sorting the color of the letters from the meaning of the word—suggesting that the visual and the semantic collided in their overly integrated brains. "There is some evidence that mixed-handers have trouble doing two things at once," explains Christman, suggesting that the nonrighties might want to avoid too much multitasking. [61]

Surely there are benefits to a more integrated brain that go beyond string instruments, or we would no longer have lefties to study—or violin prodigies to play concertos. In fact, greater communication between the cerebral hemispheres may be beneficial to many of life's tasks, as well as to the whole of mankind. The question that lies ahead is, how do left-handers, and the right-handers who share their wiring, end up with so much variation in their brains? Are they born this way, or does experience change their cerebral arrangements? And if the "standard" right-handed brain is the model, how did roughly one-third of the population end up with an *altered* brain?

5

Nature or Nurture:
What Makes a Lefty?

*Sinistrality is . . . nothing more than an expression of
infantile negativism and falls into the same category as
other well-known reactions of a similar nature, such
as contrariety in feeding and elimination, retardation
in speech and general perverseness insofar as the
infant with meager outlets can express it.*

    —ABRAHAM BLAU, 1946

*There are everywhere a certain number of individuals
who despite all their efforts, all their perseverance,
remain left-handed. For them, one must admit the
existence of an inverse organic predisposition against
which imitation and even education cannot prevail.*

    —LAUREN J. HARRIS, 2000 [1]

On the border of Scotland and England sits a fifteenth-century
castle known as Ferniehirst, built by the Scottish Kerr clan.
Entering the castle's medieval towers is like peering into a looking
glass. In most castles, the stairs leading up to the parapets wind
clockwise. At Ferniehirst, they spiral counterclockwise, so that

when you descend the stone steps, the right of you is pinned against a stone wall, while the left of you is free to wave to the ascenders on the left, or stab them willy-nilly, as the Kerrs apparently did. If you were climbing the stairs and had the misfortune to be right-handed, you'd have to use your backhand to defend yourself— that is, assuming you could manage to use anything at all, since the ascender's right arm would be trapped against a wall as well. According to Scottish lore, the castle was built by a left-hander, Andrew Kerr, forefather of the clan. Legendary for their left-handed swordfighting, the Kerrs were so deadly with their foils that no one could fight against them, according to the poem "The Raid of the Kerrs," by an Englishmen named Ettrick Shepherd from that time:

> But the Kerrs were aye the deadliest foes
> That e'er to Englishmen were known,
> For they were all bred left-handed men
> And'fence against them there was none [2]

The Kerrs survived to produce millions of lefties, for the Scottish still use the term *kerr-handed*, or *corry-fisted*, to describe all left-handers today, and *Kerr* is the thirty-first most frequent surname in Scotland today, according to the country's General Register Office. [3]

Five centuries after the Kerr clan convinced the English world they were breeding deadly sinistrals, the debate over what causes left-handedness is still alive. Are they born or made? If they're born that way, is it a normal variation—like red hair or green eyes—or a deviation, like webbed feet. If most of the human species is right-handed, what would make a person thwart convention and rely on his or her left? Adding to the confusion is this combination of facts: left-handedness tends to run in families but may not always be apparent (was Grandma switched by Great-Grandma?). Adopted children are much more likely to share the handedness of their biological parents, yet even two left-handed parents have more than a fifty-fifty chance of giving birth to a right-handed child. Complicating things further is the fact that nearly twenty percent of identical twins have different hand preferences. Clearly, no single nature versus nurture theory can explain all of these facts. [4]

Throughout history different theories have surfaced to explain how and why we humans have become such an asymmetrical species, with such a large majority tilting to the right. What happened to the lefties? In a perfectly random and balanced world, shouldn't half of us prefer our left hands and half of us our right? With advancements in science, the lefty theories have evolved and become more complex, but along the way, each theory, however biased by the times, has offered a discovery or a trace of truth, bringing us constantly closer to answering the question, is left-handedness the product of nature, nurture, or something in between?

Sword and Shield Theory

Outside of the Kerr castle, in the field battles of the early warriors, right-handed soldiers were believed to have had a life-saving advantage. Carrying their sword in the favored right hand, they could hold their protective shields with the left hand, which made it easier to cover the heart, ticking as it does from the left cavity of the chest. Even if they were stabbed, the theory holds, the dextral swordsmen were more likely to survive to churn out more righties, whereas the left-handed warriors, holding the shields on their right sides, had greater odds of taking a fatal sword to the heart.

Now, the shield, for the righty or lefty, was big and round enough to cover the heart, which is only slightly skewed to the left (one-third the mass of the heart lies in the midline of the body), so it's hard to see how the left-handed were more vulnerable. Despite the reality of shield size, this was the theory proposed during Victorian times, when the world was slowly coming to terms with the Darwinian theory of natural selection—that the strongest of a species survive to beget more of their type. In 1871, a young London physician, Philip H. Pye-Smith, proposed that the primitive condition of man was that of "perfectly symmetrical structure and ambidextrous function." Once the right-handed advantage was established by these early fighters, Pye-Smith explained, education and regular use ensured its dominance. [5]

Since the Victorian era, several holes have been thrust into the sword-and-shield theory of handedness. Anthropologists argue that evidence of right-hand dominance can be found long before there were swords and shields. In surveys of artwork and statues dating back five thousand years, ninety percent of man's ancestors were depicted using their right hands to perform manual tasks. Stone Age man, dating back thirty thousand years, appears to have been right-handed at least 75 percent of the time, according to hand tracings found in caves in France and Spain. [6]

If the left-handed did have some disadvantage in warfare, why, then, weren't all of the lefty swordsmen killed off? In fact, many left-handed warriors—outside of the Kerr family—survived to become legendary conquerors. According to the Book of Judges, Israel was freed from the tyrannical domination of King Egon of Moab by a left-handed swordsman named Ehud, who managed to plunge his double-edged sword into the heavily guarded despot's stomach. Commodus, the emperor of Rome during the *Gladiator* era was said to be left-handed. So were Julius Caesar, Charlemagne, and Alexander the Great. It would seem that many left-handed warriors not only survived but thrived. [7]

The Deception of
The Family Tree

For centuries, scholars have relied on large, well-documented families for evidence of biological traits. The British royal family provided a high-profile family tree to monitor, and several of its most chronicled members displayed left-handed leanings, despite their own efforts to militate against the trait. Queen Victoria, the stern moralist who reigned over an entire era of righteousness, may have been hiding a sinister predilection, as she reportedly painted with her left hand. Her son, King George VI, also displayed strong left-handed tendencies, despite attempts by a governess to convert him. Modern royals, including Prince Charles and his son Wills, display their left-handedness in public, continuing a clear, traceable line. The royal predilection for the trait even led some to suspect that Jack the Ripper, the infamous sinistral slasher of late-nineteenth-century London, was actually Victoria's grandson, Prince Albert Edward, the Duke of Clarence and Prince Charles's uncle. [8]

The Scottish Kerr clan is also offered up as evidence that left-handedness runs in families. In fact, half a millennium after they hacked their way into international lore, a 1974 international survey found that Kerrs (and Carrs, the anglicized version of the name) are three times more likely to be left-handed in both the United Kingdom and North America. Charlemagne, the first emperor of the Holy Roman Empire, was left-handed, and he had four successive wives and six concubines. Some estimate that half of Europe could claim him as an ancestor. With all of this hard procreative work, perhaps these two men inadvertently bolstered what would otherwise have been a dwindling population of lefties. [9]

For years, critics contested the contemporary Kerr survey that shows an enormous port-handed tilt by Kerrs and Carrs of today. After all, like many surveys, it suffers from a response bias; lefties are more likely to answer questions about a trait that they possess. Then, in the mid-1990s, a new theory arose: two British scholars, Walter Bodmer and Robin McKie, speculated that Andrew Kerr might have been *coaching* left-handedness. In their treatise on the Human Genome Project, *The Book of Man,* Bodmer and McKie revealed that Andrew Kerr's armed servants, who by custom took the family name, were also left-handed. Realizing the advantage it conferred, the patriarch encouraged all of his underlings to wield their swords and axes with their left hands—including his son and grandsons, who presumably did the same with their offspring. This opened the door to the possibility that the Kerr-fisted were *made* and not born. [10]

Nurture Theory: Lefties Are Made . . . or Unmade

Evidence is abundant that cultural forces play a role in determining the number of lefties a country produces—at least those who can be observed as lefties, or will admit to the predilection in surveys. The cross-cultural studies have revealed that where and when you were raised can have a powerful impact on which hand you use—especially when it comes to writing.

A century ago, only about two percent of North Americans were left-handed, as compared to approximately twelve percent today. Yet in modern-day Taiwan and Korea, the rate of left-handedness remains below five percent. Even in the United States, Chinese Americans, adults who attended Catholic schools, and older generations of Americans raised in the strict conformist years before World War II are much less likely to be left-handed. [11]

The social forces operating against left-handedness in Eastern societies have led scholars to concentrate on Western cultures when searching for what causes handedness. The less pressure there is on individuals to use the right hand, the "truer" a picture of left-handedness emerges, and the highest figures for left-handedness are likely to give the most realistic idea of how many left-handers there are in a population. [12]

Cross-cultural comparisons do, however, tell us two important things: that closet lefties can get by fairly well as make-believe righties, and that despite attempts to teach or coerce them otherwise, many people persist in using the left hand.

> *"There are everywhere a certain number of individuals who despite all their efforts, all their perseverance, remain left-handed,"* explains Lauren J. Harris, professor of psychology at Michigan State. *"For them, one must admit the existence of an inverse organic predisposition against which imitation and even education cannot prevail."* [13]

Although several scholars have challenged the idea of an inherent disposition for the trait, the reality is that any purely environmental theory of handedness has to explain why some siblings are left-handed while others are not, why adopted children are more likely to share the same limb leanings as their birth parents, or why kids share the handedness of their missing biological parents more than that of their step-parents. Though many researchers have tried to devise nature-based theories that incorporate these facts, only a few have caught on.

The "Flawed Righty" Theory

If there is in fact a natural predisposition for handedness, there is still the question of just how "natural" it is. One theory suggests that every left-hander was once a twin whose mirror-image mate did not make it through gestation. Another argues that the Babe Ruths of the world were once fetuses whose left arms faced outward in the womb, giving them more space to poke and prod their mother's belly—and develop an early southpaw pitching arm. These theories have been largely dismissed, but among the "in utero" explanations of left-handedness that have been tossed around over the years, one seems to prevail, despite the weight of evidence against it. According to this hypothesis, Mother Nature plays a role, just not in any genetically preordained way. Officially known as the "pathological left-handedness" theory, or PLH, it is based on a belief that unstable in utero conditions and birth complications convert normal brains into aberrant left-handed brains, switching motor skills dominance to make left what was supposed to be right. [14]

What could be described as the "weakest link" theory of left-handedness, PLH was the culmination of a stream of reports that appeared throughout the twentieth century suggesting a higher-than-usual incidence of disorders among the left-handed. Essentially, the sinistrals seemed to turn up everywhere you didn't want to be: the trait appeared to be more common among the developmentally disabled, autistics, dyslexics, epileptics, hyperactives, and the generally klutzy. We've heard about studies that linked left-handedness to psychiatric and social problems such as alcoholism, schizophrenia, depression, and criminality. Meanwhile, others claimed that left-handers tended to mature more slowly, end up shorter than average, and suffer from more accidents. [15]

Only a few of these reports were ever replicated, many were refuted outright, and all of them used varying definitions of handedness or "base rates" for the lefty population. Yet inevitably, some scientists began to connect the dots, and by the early 1980s, enough evidence was collected to introduce the "pathological" theory, based on the assumption that left-handers are, in effect, brain-damaged righties. Fetuses destined for normalcy were believed to be waylaid by minor "insults" to the left hemisphere of the brain during pregnancy or birth. The neuroscientists behind the theory proposed that damage to the left side of the fetal brain, which is believed to be more vulnerable, might weaken nerve pathways on the body's right side, causing otherwise natural right-handers to use their left hands.

The theory was popularized in the 1992 best seller, *The Left-Hander Syndrome*, which cited several of the studies linking lefties to health problems, and concluded that those who possess the trait suffer from an overall "decreased survival fitness" and shorter life spans, which accounts for the low levels of lefties in the population compared to righties. One component of the pathological theory focused on anoxia, or oxygen deficiency, during pregnancy or birth, which was believed to damage only the left-hemisphere brain cells, shuffling control of motor skills, and possibly language, to the right hemisphere. Introduced in the mid-1980s by Canadian psychologist Paul Bakan, the theory was based on a controversial study of monkeys, and several reports have subsequently refuted the findings. As UCLA neurologist Dan Geschwind explains it, anoxia is not "focal" or location-specific. If you are deprived of oxygen at birth, your whole brain is affected, not just the region that controls motor dominance. Most cases of anoxia result in some form of developmental retardation. While the developmentally disabled are less likely to acquire specialized manual skills—to either the left or the right—the link between retardation and left-handedness is weak. "If there is oxygen deficiency, that has to be documented," explains Geschwind. "And in twin studies, we would find that the left-handed twin would have an IQ that was considerably lower than the right-handed twin. And we simply don't find that." [16]

Another suspected culprit that contributed to these pathological theories was the role of sex hormones in the womb. In the early 1980s, Harvard researchers Norman Geschwind (Dan Geschwind's older-generation cousin) and Albert Galaburda discovered that left-handers seem particularly prone to certain immune system disorders—particularly Hashimoto's thyroiditis (viral thyroid infection), myasthenia gravis (a paralytic muscle disorder), and ulcerative colitis (an inflammatory bowel disease). They also appeared to be at higher risk for autism and dyslexia. In addition, data compiled from a range of handedness surveys suggested that boys were more likely to be left-handed than were girls. Separate research studies had found that boys were more likely to suffer from immune disorders, autism, and dyslexia. (Interestingly, the "survival fitness" of boys was not called into question, but that of left-handers was.) [17]

Seeking the common denominator, the researchers homed in on what appeared to be a solid suspect: prenatal hormones, particularly the male sex hormone, testosterone. While testosterone is actually present in both males and females throughout their lives, it is generally assumed to be a male hormone since it is present in males in much higher concentration. In male fetuses, higher testosterone levels help to guide male gender differentiation and prime specific brain pathways. Geschwind and Galaburda hypothesized that abnormally elevated levels of testosterone at specific stages of prenatal development—which could result from maternal stress, alcohol intake, excess body fat, or physical activity—might slow the growth of the more sensitive left hemisphere of the brain, forcing the right hemisphere to compensate. Corresponding regions of the right hemisphere, such as motor skills or language, would thereby develop more quickly, leading to left-hand dominance. [18]

Deepening their curiosity was the fact that excess testosterone had also been linked to immune disorders, since the hormone can slow growth of the thymus gland, which is responsible for certain immune system defenses involving white blood cells. In their eyes, the exposure of male fetuses to above-normal amounts of testosterone seemed to work as an explanation for some puzzling aspects of left-handedness: the preponderance of males over females and the studies that linked the trait to immune disorders. [19]

Geschwind and Galaburda concluded that testosterone was the likely cause of left-handedness and its related pathologies. Their theory was embraced in the late 1980s, for it seemed to braid together these previously unrelated and unexplained correlations. And it was so widely accepted by the early 1990s that it was used to link then-president George Bush's overactive thyroid condition to his left-handedness. [20]

* * *

Throughout the 1990s, the role of testosterone as culprit and common denominator was largely discounted. A study released in 1995 found that kids who experienced higher levels of in utero testosterone, based on studies of their amniotic fluid from fifteen years earlier, were actually more likely to be right-handed, not left-handed. A 1996 analysis of handedness and gender-related physiological characteristics thought to be associated with testosterone found no differences between "dextral" and "sinistral" females, including average breast size, age at the onset of menstruation, and normal menstrual cycles. [21]

Marian Annett, a professor of neuroscience at the University of Leicester in England, has been a leading researcher in the handedness field since the early 1970s. She, too, dismisses the role of testosterone. "Hormones influence rates of overall growth, but it is not clear that they could affect the (brain hemispheres) differently." Some suggest that if the hormone was a driving force behind the trait, the gender gap would be more like a gigantic gulf. "Left-handedness is only slightly more common in men than women, and the testosterone difference prenatally between males and females is huge," explains Jerre Levy, professor of psychology at the University of Chicago. Others argue that the gender difference, which has not been significant across all studies, can be explained by cultural variables. [22]

The Gender Gap

Surveys of lefties consistently find slightly more men than women with the trait, but some have suggested that the gap is small enough to be chalked up to social factors. At least part of the difference is a reflection of how men and women respond to surveys. "Men for whatever reason are more likely to adopt and to report extreme categories," explains Harris. "Women are more moderate in their responses." As a result, men appear more at the extreme ends of the handedness continuum, and women are more likely to show up in the middle. [23]

Interestingly, more women than men also report experiencing pressure to switch handedness. This may stem from boys and girls reacting differently to the pressures of conformity and the desire to please teachers and parents. In studies, boys tend to be less responsive to training and less interested in pleasing adults, while girls are more oriented toward adult figures, explains Harris. "There is a legitimate argument to be made that fewer females become manifest left-handers because they are more responsive to direct and indirect social pressures," adds Michael Peters. "The slightly higher percentage of male lefties than female lefties matches surveys that examine pressure to change handwriting and successful switching of handwriting." [24]

Even without external pressures, young girls may struggle harder to emulate and model themselves after the norm. They may be more interested in pleasing adults and teachers, or they may be more capable of pleasing. When children are very young, girls tend to have faster fine-motor skills than boys, Harris explains, a reflection of their earlier maturation. This might make it easier for them to be switched. "The slight preponderance of males over females in Western countries could well be due to subtle differences in which pressure comes to bear on females and on their greater responsiveness to social factors," Peters concludes. [25]

Pathologies of Pathological Theory

After ruling out testosterone and oxygen deficiency, scientists eventually came to oppose the notion that left-handedness is pathological in the vast majority of cases. To begin with, the link between left-handedness and immune disorders fell apart under greater scrutiny. After the whirlwind of reports from the 1980s that culminated with the "left-handed syndrome" claim, British and Canadian researchers Chris McManus and Phil Bryden joined forces in 1994 to conduct a full-scale joint meta-analysis of the health-related research, reviewing eighty-nine studies involving over twenty-one thousand patients. They found no overall relationship to handedness in those with immune system–related disorders, including asthma, allergic conditions, diabetes, lupus erythematosis, and rheumatoid arthritis. (In fact, certain allergic conditions were found to be more common in righties.) More recently, in June of 2001, a study by Clare Porac at

Pennsylvania State University surveyed more than a thousand adults aged sixty-five to one hundred and found "no evidence to support the suggestion that left-handers are more likely to suffer from either major or minor health problems, including autoimmune disorders." The study reinforced similar findings from a 1998 study conducted by Porac, who was a research partner with *Left-Hander Syndrome* author Stanley Coren in the 1980s. [26]

Neuroscientists today do not rule out the possibility that some people may have been on their way to right-handedness but were diverted by some act of prenatal nature. However, developmental interference to this degree is unusual and could explain only a very small percentage of the left-handed population. As Dan Geschwind explains, "It would take a significant insult or injury to the brain to switch handedness, one that would likely disrupt cognition in some way. It could not be a subtle or minor injury that lurks beneath a surface of normalcy." [27]

* * *

Picking apart the findings that connected left-handedness to problems, scientists today point out that most of the claims made about "sinistrals" and their so-called health problems were based on small, poorly designed studies, representing statistical artifacts rather than facts. "Smaller populations tend to see bigger effects, which is why some conditions may appear to be linked to left-handedness," explains Professor Ward. [28]

There is evidence of these "bigger effects" in the theory that lefties result from problematic pregnancies and births. According to a University of Toronto report, possibly as few as one percent of left-handers became so due to in utero or birth stress, while another study put the estimate at five percent at most. In the end, scientists speculate that the statistical odds of pregnancy complications creating a lefty from a destined righty are roughly the same as the odds of a natural lefty being nudged into right-handedness by minor brain insults (taking into account that there are far fewer preordained lefties to begin with). The difference is that lefties represent a much smaller group, around one-tenth the size of the righty group, so the right-handers with the same effects are less likely to surface. No one ever conducts studies of right-handers to determine what percentage has autoimmune disorders and other pathologies, so in the end, it boils down to what Ward refers to as "a numbers game." In other words, left-handers only appear to have more problems because there are fewer of them to start out with, which results in fewer of them with no problems. [29]

Other flaws in the research have to do with the constantly fluctuating "base rate" of lefties in any given population, and the conflation of data from clinical and general populations. As Porac explains, "There are hundreds of studies that have looked at the prevalence of right- versus left-handedness in a number of groups that have clinical conditions. The data are hard to interpret because different definitions of left-handedness are used." [30]

If a study defines left-handedness as the hand one writes with, the numbers can vary from as high as fifteen percent to as low as two to three percent, depending on the cultural origins and age of the population studied. If left-handedness is defined more broadly, the numbers spike upward, yet comparison studies of autism, attention deficit disorder, stuttering, or even bed-wetting often rely on different definitions and therefore different data. [31]

One particularly misleading clinical group is that of the intellectually challenged. As the level of cognitive ability goes down, the standards normally applied to hand performance no longer apply. "The average person has high expectations of their hands and is very sensitive to minor differences," explains Peters. For instance, when it comes to clicking and dragging a computer mouse, people prefer one hand over the other because it seems so much quicker and more accurate. Yet when Peters conducted studies of average people and computer mouse performance at the University of Guelph, he found that the differences between the hands, at least when it came to actual speed and accuracy when moving the mouse to a target, were small. The preferred hand was better, but not by much. "And yet we experience this difference as sizable," says Peters, referring to the stress and frustration we feel when attempting a familiar exercise with our unfamiliar hand. In severely intellectually challenged individuals, this high degree of sensitivity to differences in right- and left-hand performance is not very developed. Rather than being ambidextrous or weakly right- or left-handed, these individuals lack pronounced preferences because often they lack the highly specialized skills that are used to determine preferences. [32]

Do Lefties Die Younger?

For nearly two decades, left-handers have been haunted by an ominous threat, the presumption that their lives may be cut short. Even after more than a dozen studies have debunked the notion, it is still the most common question posed to this author when the topic of left-handedness is raised. "Don't they die younger?"

The answer, according to the National Institute on Aging, the United States Statistical Assessment Service, the American Academy of Actuaries, and investigators from at least four different universities in three different countries, is no. [33]

The shorter life span myth was born out of observations that fewer left-handers surface in older populations. In 1980, psychology professors Stanley Coren and Clare Porac evaluated statistics gathered from 5,147 Canadians aged eight to one hundred and found a sharp decline in the percent of left-handers with age. While thirteen percent of twenty-year-olds were left-handers, only five percent of fifty-year-olds and 0.5 percent of those eighty and over fell into the left-handed category. Suspecting that life, with its right-handed conspiracy, had taken its toll on the unsuspecting southpaws, Coren continued to investigate, studying statistics from the American Baseball Encyclopedia and surveying families of recently deceased people in a small county in

California. The most dramatic finding concluded that the life span of the left-handers was nine years shorter than that of right-handers, sixty-six years versus seventy-five. [34]

That's enough to change actuarial tables and drive up lefty life-insurance rates. It was also enough to spark a fierce debate and a decade of follow-up research, which failed to sustain the charge, or in some cases, directly refuted it. Two large-scale epidemiological studies had collected handedness data on participants as part of background information. In the Framingham Heart Study, one of the most thorough scientific investigations ever conducted, there was no increased risk of mortality for left-handers over right-handers. Support for these findings came from the National Health and Nutrition Examination Survey, conducted by a division of the United States Department of Health and Human Services, which actually found an association between left-handedness and a longer life span. "The evidence points to right- and left-handed people having the same risk of death at any age," concluded Marcel E. Salive, an epidemiologist for the National Institute of Aging. [35]

So how to account for the scarcity of senior lefties? While some scientists challenged Coren's numbers, others rejected his interpretation of the findings. The sensational premature-death claim failed to take into account that older generations contained many converted left-handers. The apparent paucity of them stemmed from the fact that they were switched when they were younger, and as a result show up as right-handers in many databases.

When switched left-handers or nonwriting activities were taken into account, as was the case with several follow-up studies, the numbers told a different story. Coren's former research partner, Clare Porac, has conducted several more recent surveys on handedness and age, all of which found no gap in the life span of south- versus northpaws. In 2000, Porac examined a population of 1,277 people over age sixty-five and indeed found fewer lefties: 6.9 percent, compared with ten percent in a general population study. In those over eighty, only three percent showed up as left-handers. Then she looked at activities other than writing. Nine percent of the over-eighty group ate with their left hands, for example, and nearly every member of this group said they'd been urged or forced to switch hands in their youth. Researchers have also found that people become increasingly right-handed with age, in part because of the need to adapt to right-handed designs (computers, tools, cooking utensils), and in part because the right hemisphere of the brain is believed to age faster than the left. [36]

To strip away all contemporary social influences, Marian Annett of the University of Leicester excavated a study from the nineteenth century that tested nearly seven thousand men across all ages for the strength of grip in each hand. There was no significant difference in the proportion of left-hand dominance by age. "The absence of age trends show there was no differential mortality for handedness in the nineteenth century," Annett concluded. "Neither was there in the twentieth century." [37]

As for the popular myth that lefties are prone to life-threatening accidents, a decade of follow-up studies in Europe and North America found no elevated risk of accidents, major or minor, among the left-handed. As Clare Porac concluded from her extensive 1998 study, "The only significant finding was the left-handed found common cutting implements less easy to use when compared with responses from right-handers." [38]

Nature Theory: Handedness Is Inherited

Perhaps the biggest challenge to the theories that lefties are simply derailed and damaged righties is this: having one left-handed parent greatly increases the odds of becoming a lefty, and having both parents left-handed increases the odds even more. While mothers and fathers can influence hand preference, parental coaxing can't explain how adopted children are more likely to share the hand preference of their birth parents.

If nurture and developmental problems are not ultimately swaying the fate of left-handers, could it be that their lopsidedness is innate, a built-in bias directed by DNA? Charles Darwin may have been interested in the answer. The man who revolutionized science and redefined humanity had a grandfather, a wife, and two sons who were left-handed. Although he never wrote about the topic at any length, in his notes and letters he stated that "handedness was heritable," according to Lauren Harris, who conducted an in-depth analysis of Darwin's writings published in *Brain and Language*. Today, 130 years after Darwin's heyday, advances in microbiology and imaging technology are offering more and more evidence that our hand preference is inherited—set in motion long before we come through the birth canal, or even before our uterine biosphere can have any effect on our development. [39]

While it is tricky to assess handedness before motor skills take shape (usually in toddlerhood), some reports have claimed that our preferences, in their earliest form, are established by the time we are born. A group of researchers in Ireland analyzed images of human fetuses in utero and found that the majority were engaged in right-handed thumb-sucking. In another study, they found that eighty-five percent of fetuses moved the right arm more than the left arm. Follow-up studies led them to conclude that a right thumb preference in the womb was correlated with right-handedness later in childhood. [40]

To truly sort out the relationship between heredity and environment, scientists look to adult twins. This is exactly what Dan Geschwind and his colleagues did at UCLA in 2002. Comparing the brain scans of large samples of identical twins against a control group of fraternal twins, they found that heredity shapes the brains of left-handed and right-handed people differently. Those with left-handed DNA inherit a tendency to develop a more symmetric brain, while those with right-handed genetics inherit a tendency to develop a more *a*symmetric brain, with a larger, more developed left hemisphere. Michael Gazzaniga, a neuroscientist at Dartmouth who has studied brain organization for more than thirty years, hailed the findings as the strongest evidence to date that left-handedness is inherited. [41]

The twin study also surprised many by revealing that identical twins could share the same brain organization but not the same handedness. This has implications for all right-handers with a left-hander in their immediate family, as they could very well share the same genotype, and therefore the same brain lateralization tendencies, as their siblings, parents, or offspring. [42]

Inheriting Chance: The Right Shift Theory

Scholars from across academic disciplines have offered several theories as to how heredity ultimately shapes hand preference. Conventional models of genetics have a difficult time predicting the handedness of children born to righty versus lefty parents. This is due to the fact that although it's now widely accepted that heredity plays a significant role, handedness does not follow the typical patterns of dominant and recessive traits.

The most widely accepted genetic model for handedness is called the right shift theory, first introduced by Marian Annett in 1972. Annett suggests that a single, hypothetical gene, possessed by most of the population and designated as R+, drives the left side of the brain to become more developed for speech machinery and fine-motor movements, which leads to a favoring of the right hand. What's unique about the theory is that the other form of the gene, R-, which is carried by left-handers, does not code for a left shift. Instead, it is indifferent to the direction of both language and motor dominance. To refresh our memory, our genes come in pairs of alleles; we receive one allele from each parent. The alleles may be identical, or they may differ from each other. In the case of Annett's theory, the two alleles on the handedness gene are R+ or R-. Because each of us has two copies of the gene, our alleles can fall into any one of three combinations: R+ R+, R- R-, or R+ R-. [43]

Annett's suggestion is that the majority of modern humans inherit at least one right-shift gene, which weights the odds in favor of right-handedness. According to her theory, the way in which this gene drives the development of the left hemisphere is to impair or "prune" cells in the right hemisphere at some sensitive period of cerebral growth. Shearing the right lobe ensures that critical functions such as speech and motor skills have no choice but to end up near each other in the blossoming left lobe. The presence of R- allows for more freedom in the brain's development. Without the neurological nudge of R+, the right hemisphere is free to grow in those who inherit at least one R-. Language and motor skills are no longer forced to shuffle off together, so they are allowed to distribute more randomly.

Right-handers receive at least one and sometimes two copies of the R+ gene. Annett estimates that nearly half of the population, forty-nine percent, inherits a copy of R+ from just one parent, resulting in a mild tendency toward right-handedness. The single right-shift gene causes weak impairment of right-hemisphere functioning—enough to dampen their left hand's skill development, but not enough to squash it completely. Following the same model, those who receive two right-shift genes, R+ R+, an estimated one-third of the population, are likely to end up with weak left-hand performance and a strong tendency to right-handedness—the result of extra pruning of the right hemisphere.

This leaves approximately twenty percent of the population to receive an R- from each parent. In this group, the direction of handedness becomes a crapshoot, a flip of a coin. About half of these "double negatives" should become right-handed, and half should become left-handed. Yet due to chance factors and the right-handed world, Annett estimates that only about one-third of the double-negative group (six to seven percent of the population) show up as strongly left-handed. The other two-thirds or more (twelve to thirteen percent) would likely write right-handed but show other significant left-handed tendencies, such as throwing, hammering, or striking a match. In R- R- cases, it still remains unclear which specific factors are responsible for determining handedness, but Annett suggests that the path is determined by changes that occur early in our fetal development, as the right-hemisphere synapses are shaped. Others suggest that handedness is determined in the earliest years of our development, as we shall see ahead. [44]

Of course, there is the question of why we have two forms of the gene to begin with. Annett speculates that the R- serves to prevent us from becoming too committed to the left hemisphere, which in itself may not be to our advantage. Her hypothesis here is one that is known as balanced polymorphism and is best illustrated by looking at the gene for sickle-cell anemia. Under normal circumstances the blood disorder occurs only in people who inherit the gene from both parents. Yet in areas where malaria is present, a person who inherits just one copy of the sickle-cell allele actually has a survival advantage over someone with two normal alleles, as a single dose of the gene mutation works like an immunization against malaria. Sickle cell anemia is therefore the cost some individuals have to pay so that others with mixed genes have a better chance to survive malaria—at least according to evolutionary theory. If we stretch the analogy, the existence of the R- gene prevents humans from being overly reliant on the left hemisphere at the expense of the right. According to Annett's model, it is best to have the mixed R+ R- combination, while either R+ R+ or R- R- is less desirable. The mixed R+ R- combination is inherited by the most people, which means a smaller number will have the "risk" of being R+ R+ or R- R-. Following Annett's theory, R+ R+ or R- R- individuals are at a disadvantage when it comes to cognitive abilities, but it is at this stage in her theory where many scholars begin to disagree. [45]

Though many scientists take issue with the theory's link to cognitive skills, the genetic model behind Annett's theory has proven to be very successful at predicting the handedness of offspring and explaining the complexities of how the trait is inherited. "The strength of Annett's genetic model is that it can explain observations that are extremely hard to account for by traditional models," explains Peters. Her theory generates a solution for one aspect of inheritance that has puzzled researchers for years: that identical twins can end up with different hand preferences. According to Annett's model, this makes perfect sense. If the twins both inherit either R+ R- or R- R-, there will be a large chance element playing a part in determining the final outcome of their hand preferences.

Another quandary is that left-handed parents have a greater likelihood of having lefty children than right-handed parents do, but still the majority of their children will be right-handed. With Annett's model, if both parents are R- R-, we can expect half of their children to be left-handed by chance alone. However, because we live in a word of right-handers, environmental pressures will undoubtedly lead to well over half of these children becoming righties. Exactly how many depends on the estimates of the frequency of the R+ and R- genes in the population, a statistic that researchers can only guess at.

So left-handers do not inherit their hand preference per se. What they inherit is a lack of neurological bias that nudges the left hemisphere to become dominant for language and motor skills. This means that they will not share the "standard" brain arrangement of right-handers, and one left-hander may be quite different from the next. The strongest support for this prediction comes from Dan Geschwind's 2002 twins study. As expected, the brains of identical right-handed twins were very similar in size and structure. The left brain hemisphere was larger in volume size, an indicator of language dominance. This changed when one member of the twin pair was left-handed; in this case, the brains of both members were more symmetric in size. The left and right hemispheres were more equal in volume, suggesting that Broca's and Wernicke's areas were less restricted to the left hemisphere. As Geschwind explains, when one member of a twin pair is left-handed, both the twins "have a tendency toward symmetry, and whether they end up left- or right-handed, their brains look the same. That's a new finding." [46]

Another important implication of these findings supported Annett's theory of handedness: in left-handed twins, genes contributed less to the particular configuration of the cortex than in right-handed twins. "That is exactly what one would expect from left handers as a group, because chance plays more of a role in their inter-person variations than is the case for right-handers," adds Peters. Exactly as Annett predicted, the layout of the cerebral cortex in identical twins who are both left-handed is not as strongly under genetic influence as it is in right-handed twin pairs. The presence of R- grants the twins who carry this gene more flexibility when it comes to the blueprint for the development of their cortex. [47]

Recently, Annett's R+/R- theory has once again ventured to the forefront of genetic research with the mapping of the human genome, the basic structure and content of our DNA. Now it should be only a matter of time before scientists uncover the hypothetical R+ gene. In 2003, a group of scientists at Oxford University conducted the first genomewide analysis to pinpoint key regions that may be influencing handedness. Screening 105 pairs of adult brothers in which at least one of the two was strongly left-handed, they found that relative hand skill was influenced by at least one genomic region, located on chromosome #2, which is known for many, many things, many of which are completely unrelated to handedness or even the brain. (For example, red hair, precocious male puberty, pancreatitis-associated protein, deafness, and glaucoma show up on the list. But random traits such as these tend to show up on many genes.) The region the Oxford scientists have homed in on is considered relatively large, spanning more than one hundred individual genes, and the scientists have yet to identify which specific gene or genes underlie the effect. Nonetheless, the findings offered evidence that "while the causes of handedness variation are complex, there is certainly at least one genetic influence on this trait," according to Clyde Francks, an Oxford geneticist who led the genome study. [48]

Multiple Genes, Same Outcome

The right-shift model proposed by Annett was based on a hypothetic gene, the R+. Yet considering that even presumably simple traits such as whether you can curl your tongue are thought to be the result of multiple genes, we can be sure that nothing is that simple about brain lateralization and handedness. Given the complexities of brain-body organization, and the variations in how left-handers' brains are organized, many neuroscientists and geneticists believe that handedness must be caused by two or more genes, and some see it more as a chain reaction of genetic events.

Molecular biologists have suggested that a whole family of genes may be activated to determine handedness, an unfolding sequence of cascading influences, and that the source of left-right asymmetry can be traced back to the very first gene and to that first cell in which that gene goes to work. "The direction in which the cilia on cells beat may well depend on the asymmetry of molecules," explains Jerre Levy of the University of Chicago. And it could be that the direction of the little cell hairs ultimately sends certain brain cells—those that lead to the development of speech and motor skills, for example—to the left or right hemispheres of our cortex. Dan Geschwind, who is both a neurologist and a geneticist, points to the connection between the lateralization of the brain and the asymmetrical development of the entire body, including the differences between our fronts and our backs. "The idea is this: the left-right axis is just another axis. And our body is very differentiated along the left-right axis, as is our nervous system." [49]

When an embryo is first developing, a series of common DNA blueprints operate like mini molecular traffic directors, lining up along complex signaling pathways, organizing early cells into organs while signaling them to migrate to the left (as is the case with the heart) or to the right (the liver). Like a molecular relay race, certain genes switch on other genes that switch on others, creating a whole pathway of genetic events. It may be that a path of genes in each species is responsible for asymmetrical development of the brain. [50]

The hypothetical right shift that occurs in righties—and is lost in lefties—begins with these early triggers, setting off a domino effect of directional flagging that occurs during different stages of development. Ultimately this leads to the asymmetrical specialization of our brains, which in turn creates a dominant motor center in one hemisphere and a corresponding dominant hand on the opposite side of the body. [51]

What is less clear in the model is why in nearly fifteen percent of the population, these early molecular signals send "motor neurons" to the right hemisphere of the brain and not the left. Geschwind theorizes that this is due to genetic mutations, or spontaneous changes in genes that either render them dysfunctional or alter their function in some way. Due to these mutations, some people inherit slightly less side bias, such as the small percentage of people whose heart is centered slightly to the right of the midline in their chests, rather than to the left. This "laterality pathway" begins very early on in development, explains Geschwind, and depending on where and when along the pathway a mutation occurs, it may alter brain laterality, organ asymmetry, or both.

Geschwind and his colleagues concluded that while genes play a strong role in determining the volume of the brain hemispheres—and the accompanying cerebral dominance—environment can also play a role. This brings us back to the missing links that might explain how identical twins can end up with the same brain asymmetry but different handedness. While genes may be dictating their brain asymmetry, it is less clear how twins with the same brain structure are ending up with opposite motor dominance.

The "developmental" explanation goes something like this: sometime between the sixth and the twentieth week of pregnancy, when the central nervous system is being formed, the geographic fate of brain cells is determined. Those mini molecular traffic directors set to work, sending some neurons to the left half of the brain, and some to the right. They line up along a route in the developing embryo, flagging juicy brain cells to the left or to the right. In one identical twin, the traffic directors may send more neurons—or a different type of neuron, one with more dense connecting fibers—to the left hemisphere of the developing brain, where they end up somewhere in the general vicinity of the motor-skills center. In the other twin, these genetic traffic directors may randomly send more "motor headquarters" neurons to the right half of the brain, increasing the likelihood of producing a left-hander.

While DNA's traffic directors may explain how a small group of identical twins ends up with different hand preferences, it does not tell us anything about how each parent's genes influence these preferences. "The problem is that in matters of handedness we cannot draw any specific conclusions about individuals," explains Michael Peters. "Predicting the most likely handedness of a person on the basis of the parents is strictly a matter of statistics and averages, which cannot tell us about the individual case."

This is apparent in the situation of Rich Cohen. A contributing editor to *Rolling Stone* magazine and the author of five books, Rich is the third right-handed child born to the dual-lefty team of Herb and Ellen Cohen. Though he has no recollection of learning to write right-handed, he had a terrible time writing as a kid and was not much of a student. "It was only when I learned to type on a computer that I was able to write fluidly," says Cohen. So can we assume that Herb and Ellen are both R- R-? Unfortunately not. They could have had any of three combinations: both of them were double negative or R- R-, both of them carried the combo R+ R-, or one of them was R- R- while the other was R+ R-. Any of these pairings could produce right- or left-handed offspring.

In addition to knowing where in the genome the handedness gene(s) are located, we also need to know how these genes influence the outcome of hand preference. We've seen that random factors can affect how genes are expressed early in the molecular stages of development, but it is also important to understand what shapes the development of handedness in early childhood. Kevin Laland of Cambridge University suggests that random events during a child's early years (a mild injury, repeated imitation of an older sibling, parental coaching) play a role in disrupting or reinforcing the underlying influence of genetic makeup, and these influences play a bigger part in those who lack the strong genetic directive of the right-shift gene. "Young kids will use both hands equally, and only later specialize," explains Laland. "Perhaps they get used to using one hand more than the other, or by chance have more success with one hand." [52]

Regardless of the influence environment has on our hand preference, the issue today is not *whether* genes make a difference, but *how much* of a difference they make. Scientists currently adhere to the idea that for some individuals (many lefties), genes make less of a difference in the determination of hand preference than for others (most of the right-handers). Despite this, "nature" is still considered the most powerful influence on handedness, and given the evidence that our brain and motor system begin to organize themselves early in our development, scholars from several disciplines now urge parents and teachers to avoid "nurturing" any handedness outcomes in their kids. [53]

Social Darwinism and the Left-Handed Gene Pool

While nature may play the strongest role in determining handedness, there is still the question of how different populations end up with starkly different numbers of left-handers. In most Western countries, the rates of lefties are relatively consistent across race, gender, and socioeconomic status. Yet their numbers drop off in countries that place a heavier weight on conformity, as we've seen in previous chapters. The question is how the "nurture" of social values can affect the presence of a natural trait.

Part of the international culture gap can be explained by limitations in research and reporting, the difficulty in gathering representative samples and honest answers from citizens who do not wish to reveal their differences. But research variables alone cannot account for those skewed results. While early parental involvement can influence some aspects of a child's handedness, social forces can also impact the frequency of a genetic trait. In his book *Right Hand, Left Hand*, University College London professor Chris McManus theorizes that waning discrimination against left-handedness in the twentieth century may have helped spur on an increase in the chance genes that lead to the trait, resulting in greater numbers of lefties alive today. The weight of prejudice, particularly in Victorian England, was strong enough to have interfered with the mating prospects of the left-handed, limiting their reproductive potential and resulting in fewer offspring, McManus speculates. That may seem like a stretch, but considering that left-handedness could be used as grounds for divorce in Japan as recently as the twentieth century, it might explain some of the lower frequency of the trait in that country. [54]

"Indirect social pressure is a far more insidious force than direct pressure," McManus explains. "A little gentle gossip, a few carefully aimed barbs, a sneer or tow at the right moment, and people find themselves ostracized, isolated, spurned, and, as a result, possibly even no longer considered as a possible sexual partner. Within small, pre-modern, pre-technological societies, the effects of such pressure could be quite large, particularly in a world where most people married partners who had grown up within the neighbouring ten or twenty miles. Indirect social pressure does not stop people altogether from finding a partner or having children, but it might delay the process, and the later one starts to have children the fewer one is likely to have." [55]

In 1992, McManus and a Canadian colleague, Phil Bryden, reviewed data from historical studies and found that in the first half of the twentieth century, right-handers had significantly more offspring than did left-handers, and more offspring is considered, from an evolutionary perspective, proof of the success of certain genes. Before 1955, two right-handed parents averaged 3.1 children, while the presence of a left-handed parent decreased family size down to 2.69 kids per family, a significant drop-off. If both parents were left-handed, the average number of children produced was down to 2.32. After 1955, the trend reversed: the presence of a left-handed parent actually increased the average size of the family. One lefty and one righty produced 2.6 children, two lefties produced an average of 2.57, while two right-handers had only 2.49 children—a statistically significant difference.

"Presumably, if there is evidence that this process was at work at the beginning of the twentieth century in the west, it may also be acting in the less developed world at the beginning of the twenty-first century," McManus concludes. [56]

* * *

Rapid advances in research have shifted the center of gravity, and it now appears that there are different causes for the same outcome— left-handedness —only a very small subset of which could be chalked up to "pathological" factors. Today, the weight of evidence points to heredity as the driving force behind handedness. Lefties carry their own genetic combinations, and these genes are expressed in a way that's unique to each individual, allowing for different variations within the southpaw population.

The Scottish Kerr clan might have been encouraging left-handedness, and the British royal family might have been discouraging it, but both ended up producing more left-handers. It may have taken a few centuries, but science is finally proving that the left-handed are not just right-handers gone amok. They are one of Darwin's variations, important enough to have survived and thrived throughout human evolution—despite so many right-handed staircases, wet nurses, and shillelaghs smacking them into submission. The question is how—and why?

6

The Tipping Point:
Evolution and Handedness

*Why weren't the left-handed killed off? Were they
wicked and perverse people who refused to listen to
the good prehistoric surgeon-general, when he told
them to carry the shield on the left, and who, through
some lapse of justice, escaped their deserts?*

—THOMAS DWIGHT, 1891 [1]

Augusta is a southern belle of a chimp who does not like to be
ignored. If you walk by her, she'll throw something at you: a
lump of banana, a plastic tube, feces. Her wind-up is strictly Little
League, but she has been known to hit some targets, including the
face of Bill Hopkins, her sometimes coach. When Augusta feels
threatened, which happens when she is greeted by oversized upright

men, usually newcomers, she throws even harder, once even ensuring that her honey supply would not be cut off again. Augusta is, of course, a lefty.

Or is she? Consensus may be building that left-handedness is a natural trait, but scholars are still at odds when it comes to how far back in our evolutionary history the trait first appeared. On one end are those who believe early man was the first to pick sides. On the other are those who trace the predilection back to something as lowly as the fiddler crab. Both schools have to explain what benefit a species gets from "specializing" on one side, and how modern-day humans ended up overwhelmingly biased toward their right sides. Was left-handedness some sort of mutation, an accident of nature? Or is it more like the appendix, an obsolete vestige left over from some prehistoric purpose?

Perhaps, like our opposable thumbs, left-handedness has some purpose that kept it alive as our species evolved. The big evolutionary question that surrounds handedness is whether man is alone in his collective preference for the right hand, and, if not, how far down the food chain the proclivity goes.

Before we can understand how the fabled southpaw came to be, it helps to unravel when and why we decided to pick sides in the first place. For most of the twentieth century, it was generally accepted that the tendency to favor one hand over the other was a distinctly human characteristic, related to our specialized brains and our unique ability to speak, understand language, and play charades. It was seen as one of the traits that set us apart as a species. On this assumption, the great apes before us were ambidextral indifferent-handers. After all, what need did they have for the nimble skills of a well-practiced hand when they were still relying on all of their hairy knuckles to help their big hairy feet get around? Augusta, the temperamental southpaw, still relies on her hairy knuckles to get around, but once she stands up, her knuckles serve other purposes: rummaging for termites in a tree or peanut butter in a tube, shoveling these edibles into her mouth, or grasping a banana lump to throw at an annoying human.

What occurred between Augusta's ancestry and our own is a topic of great debate. In one camp are scholars who subscribe to a kind of "big bang" theory of human evolution, which suggests that something sudden and dramatic occurred during evolution that made humans unique and incomparable to other species. In this camp you'll find linguistics professor Noam Chomsky and other evolutionary theorists who point to human language, which requires brain lateralization similar to that required for handedness, as the big leap separating man from all other species. The other camp argues that most, if not all, human traits have precursors in the species that came before them. Regardless of the trait's origin, both camps agree that our specialized motor skills are tied to our highly evolved human brains. [2]

Throwing Rocks, Nailing Supper

At some point between the chimp and the hominid, man's earliest ancestor, throwing became more than just a way to express anger and mask fear. On the savanna, the ability to aim and hit a target with some force was one way to bring home supper. Most likely, it began with a simple rock, an effective weapon when thrown at small birds or scurrying prehistoric groundhogs. From this humble origin, early hominids learned that a single-handed throw would provide greater speed and accuracy when it came to hitting a moving target. Evidence that early man was throwing rocks to hunt can be found 1.5 million years ago in the form of Acheulean hand axes, flat, aerodynamic rocks rounded into tear drops with sharp edges all around. As Robert Ornstein speculates in *The Evolution of Consciousness*, these handleless weapons could fit in the palm of the hand, and tossed over the head, they were well suited for hunting birds. [3]

In the 1870s, Charles Darwin spent several months on the islands south of the Straits of Magellan, observing a primitive tribe called the Fuegians. Isolated from the developments of Western civilization, the tribe lived as prehistoric man did: throwing rocks and spears and arrows to kill birds, hunt prey, and defend their offspring. "To throw a stone with as true an aim as can a Fuegian in defending himself, or in killing birds, requires the most consummate perfection in the correlated action of the muscles of the hand, arms, and shoulder, not to mention a fine sense of touch," Darwin remarked in *The Descent of Man*. [4]

This need to execute high-speed, sharp-aim throwing to kill prey became a driving force of evolution, laying the groundwork for the neural wiring that would eventually orchestrate other, finer-tuned correlated actions such as gesturing and, eventually, speaking. William Calvin, a neurophysiologist at the University of Washington and author of *A Brief History of the Mind*, believes that throwing put the brain on an evolutionary fast track toward specialization in one hemisphere, though it was less a single invention or "big bang" than a long climb. Those early ancestors who first developed a weapons-grade throwing arm would have been the "fittest" members of the species, bringing home more dead birds, staving off more predators, and surviving to produce more offspring. "Judging from the delight which infants and young children take in throwing and hammering, and from right-handedness being strongest for such tasks, the neural machinery (associated with these tasks) must have been under substantial selection at some time in evolution—probably in hominid evolution, as these features are far more pronounced in humans than chimpanzees," explains Calvin. [5]

"Selection" refers to Darwin's theories of how we evolved. Natural selection is based on the idea that certain traits are passed on because they help a species survive. While natural selection can explain how sharp teeth and claws evolved in carnivores, it can't explain why the human brain became three times the size of the chimp brain, developing all sorts of extravagant capacities such as artistic and musical abilities that were never needed for survival. This is where Darwin's theory of sexual selection comes in. It wasn't enough to merely survive; early man and woman had to be sexually desirable to produce offspring. (Any behemoth can club a bird, after all, but does he make the females of the species scamper and hide after witnessing this brutality?) Traits are sexually selected when they make a candidate more attractive to the opposite sex, providing them with a greater number of opportunities to reproduce, which in turn makes their traits more prevalent in the species.

While Augusta's feces-throwing abilities are not likely winning her lots of dates, Derek Jeter's precise midair throws have earned him plenty of female attention. The question is whether human-style coaching—and the promise of a ticker-tape parade—could help Augusta knock out some birds and win over at least one picky, highly evolved chimp.

The ability to consistently aim and hit a target at damaging speeds requires very careful coordination and precise timing, which requires a complex tangle of neurons. As demand for precision and accuracy in throwing increased among early hominids, so did the numbers of "timing" neurons in their developing brains. This neuron-rich throwing circuitry that developed in the evolving human brain helped to pave pathways for another very important advancement: the production and reception of language. Today this is best illustrated in brain scans, which allow scientists to see how language and precise movements intertwine. When a person is asked to mimic a series of facial expressions that accompany communication and require precision movements, blood flows to an area of the brain that is also used for discriminating between sounds associated with units of speech. This can also be seen with stroke victims who lose their ability to speak and have problems with the precise sequencing of hand movements. Perhaps the best evidence of the relationship between language and the fine motor movements of the mouth and hands can be witnessed in young kids. According to Calvin, "Children learning to write often twist about their tongues as their fingers move, in a ridiculous fashion. It looks as if that broad smear of wiring may sometimes cause simultaneous activation of rather different muscle groups." [6]

Calvin believes the path that led to this intricate interconnected wiring in humans—which includes specialized brain circuits designed for synchronized and carefully timed movements—was originally selected because of hominid throwing success. This overlap in the wiring might have allowed language circuits to contribute to precision motor movements and vice versa. So the sequencing of sounds to make words would have come about through use of the same timing circuits that developed for precision movements, such as throwing a stone or wresting a branch from a tree. [7]

Chiseling Tools, Choosing Sides

After knocking out small prey with their fastballs, the early hominids had to figure out how to get to the meat within. A sharper stone would have been an effective tool, but finding rocks sharp enough to pierce hides would have been difficult. At some point after the hands were freed from the clumsy duties of locomotion, hominids began chipping stones down to create sharper edges that could be used to kill prey, cut open hides, and extract meat—providing them with heaping portions of protein. These early tools could also be utilized to shape branches into useful tools/implements and reeds into strips for weaving. This ability to create a practical cutter tool from a lumpy, amorphous rock was a significant advancement in the evolution of prehistoric man, and it was no simple feat. To end up with consistently sharp edges, the toolmaker had to select "chiseling" rocks with a shape capable of whittling down another stone at an optimal angle. One hand had to become the steadier of the object, while the other had to act as the action hand, chipping and chiseling and sculpting. From the beginning, it helped if you honed the skills of one over the other. Nicholas Toth, an archaeologist at University of California Berkeley's Institute of Human

Origins, discovered several of these stone tools at African sites, and he spent five years attempting to make and use similar models—cutting through bones and hides and carving wood—in order to understand how the early hominids lived. Toth determined that a slight majority of these early stone tools, approximately fifty-six percent, were chiseled with the right hand, which would date the first "right shift" back to 1.5 million years ago. (In Toth's estimation, the percentage of righties was probably larger, but the shape of the chiseling rock would often dictate which hand had to be used.) [8]

Michael Corballis, professor of psychology at University of Auckland in New Zealand, suggests that these fine-motor skills necessary for toolmaking were most likely introduced into the gene pool around 2 million years ago, possibly as a mutation in one individual. The primary role of this genetic force would have been to nudge more specialized neurons into one hemisphere of the brain, building up enough of a critical mass to create a control center capable of directing the intricate movements that were essential for creating and using the primordial hammer. Adding uniformity to the tool would most certainly help the tribe, since more people would be capable of using it, making the tribe more productive. This uniformity would have also provided a model for young tool users to follow, which would lend itself to easily instructing a tribe's youth. So a cultural offshoot of the early acquisition of motor skills was that it gave way to a practical favoring of one hand. [9]

Sharp objects may have helped to kill prey and build better huts, and surely these accomplishments helped with survival and made the toolmakers more "eligible" members of the tribe (provided they stopped scraping and spearing and jimmying hut contraptions long enough to mate). But there is still the question of how prehistoric man made the grand leap from slicing with rocks to speaking with words. And what happened to the poor cave people who could only chisel rocks with their left hands? Did most of them die out?

Gesturing, Signing, and the First Pickup Line

Corballis and other scholars, including Marian Annett, believe that something higher—the need to communicate—ultimately drove the human brain to become specialized for language. The single-handed motor skills required for signaling and gesturing developed in sync with another finely tuned motor ability—that of vocalization. This explains how speech and handedness ended up side by side in the same hemisphere of the standard human brain. [10]

It would have started with simple pointing, perhaps to indicate the direction of home base or a meaty carcass. If danger lurked, arms and hands were used to mime the behavior of a lion, or signal to the others to move in the direction opposite the predator. Like Lassie when she wanted someone to follow her, a few high-pitched sounds might have been produced for added urgency—even grunts and shrieks require carefully coordinated movements of the mouth, jaw, tongue, and vocal cords. It makes sense, given that we have one mouth, that vocalization would be controlled from a single "orchestration" center in the brain. [11]

"I think speech is simply half-swallowed gesture," says Corballis. "When the hominids first diverged from the African apes, they were better equipped to develop communications systems based on the hands, as they already had a good deal of control and sophistication in handling tools." The fine-motor skills developed for tool use paved the way for more manual gestures, a form of early communication that became linked to vocal language. "It's quite possible that what set humans apart was that speech began from gestures, which would explain an indirect association with handedness." The sophistication of modern-day sign language provides evidence of the importance of hand gestures in communication. Studies have found that both the left-handed and the right-handed gesture more with their dominant hand when they are speaking spontaneously. (Interestingly, the left-handed are also more likely to gesture using both hands, and they are also more likely to have developed speech centers in both hemispheres of the brain.) [12]

The early hominid with the mutation that allowed him or her to turn gestures and grunts into a mating call would've captivated the opposite sex, according to the sexual selection theory. "It was much more advantageous for Homo erectus to be able to speak for the purposes of mating," offers Oxford University psychologist Tim Crow. So despite the modern-day myth of the sexual appeal of the "strong, silent type," among our ancestors it was the early talkers, or crooners, who had the edge in the dating market. Darwin himself suggested in *Descent* that human language and music evolved, like birdsong, for courtship. Therefore the creation of a specialized motor system, which allowed for both gesture and vocalization, evolved, at least in part, through sexual selection. Most likely by chance, fine-motor capabilities were developing in the left hemisphere for the majority of these early evolvers, so the specialized hand increasingly became the right. [13]

Through this lens, it would seem that evolution has passed by the clumsy and mute left-handers. If they lacked this neurological nudge in the left hemisphere, how did they stay in the mating game? Despite all of the forces working against them, the non-right-handed managed to hold their ground. By the time the Ice Age arrived 30,000 years ago, roughly twenty-three percent of cavemen remained left-handed (down from forty-four percent 1.5 million years ago, according to studies of prehistoric handprints found in caves across France and Spain). While it would appear that the lefties were fading, they were still in the race. [14]

Planet of the Right-Handed Apes

Before we can understand how the left-handed members of the species managed, we need to look at where and why handedness shows up in other species. After all, if the rest of the animal kingdom lacks the ability to speak, they must be missing the neurological nudge that drove humans onto the higher plane. Yet their mating calls, nut-cracking skills, and banana tosses must mean something.

The line between "hand indifference" and a specieswide preference for one side is considered by some to be among the last sacred barriers separating animal from man. As recently as the early 1980s, conventional wisdom in the science community held that animals did not pick sides. Our closest cousins in the primate family demonstrated both righty and lefty tendencies; however, as a species they were believed to be evenly split among the two identities. But if animals could be proven to show hand preference, then handedness in humans would transcend our intraspecies evolution and connect us to a longer tradition in the animal kingdom.

In 1980 a few pioneering primatologists began to report collective preferences among certain species. In Seattle, Patricia Kuhl at the University of Washington found that the owl monkeys in her laboratory consistently used their right hands to perform well-practiced, precise manipulations such as turning a knob to open a container full of food. Bill Hopkins of the Yerkes Regional Primate Center in Atlanta discovered that a disproportionate sixty-seven percent of chimps in his lab were right-handed for fine-motor tasks, particularly when it came to feeding themselves—plucking raisins from a pile or extracting peanut butter from a plastic tube. Roughly the same number threw right-handed—usually wet chow or feces, which they tossed out to get attention or intimidate a stranger. Similarly, a majority of aye-ayes, nocturnal "bat monkeys" known for their large middle fingers, were observed digging for insects in tree branches most often with their right digits. [15]

These new findings opened a Pandora's box. Suddenly our lofty specieswide tendency to prefer the same hand was being traced back as far as creatures that lived tens of millions of years ago. And what need would they have for a specialty hand or a concentrated brain when they had no use for forks or sweet nothings? Even more perplexing, why would a significant majority of the monkeys choose the same hand, in the absence of a right-nudging gene, nuns, computer mice, and student desks to force them to do so?

Some scientists challenged the new primate-handedness findings, claiming they were based on artificial tasks or could have resulted from chimps mimicking researchers. Lending credence to this argument were Linda Marchant and William McGrew of Miami University in Ohio, who observed wild chimps in Tanzania in 1995 and concluded that they shared no hand preferences in their daily behaviors. Their peers, however, argued that the lab is the only place to truly test for laterality. "Observing patterns in animals in the wild is nearly impossible," argues Peter MacNeilage, a professor of psychology at the University of Texas, Austin. "Their behavior can vary dramatically depending on variables such as where they are in a tree." In the lab, scientists can control the circumstances so that the chimp has to make a choice, and patterns can be observed. They can also design tasks that force the animals to subordinate one hand over the other, such as grabbing a pipe with one hand and using the nimble fingers of the other to extract the peanut butter hidden within. The lab also makes it easier to get the numbers needed to draw real conclusions. In Hopkins's lab, there are 209 chimps in a study, some of which are raised by their chimp mothers, and some of which have human "parents." Interestingly, chimps raised by humans are more likely to share the handedness of their furrier birth mothers. As for the mimicking, Hopkins points out that no humans in the lab are throwing things, least of all feces. So if Augusta the chimp feels compelled to throw some at Hopkins's nose, this is out of instinct and not a desire to get recruited into the major leagues. [16]

Since 1990, a few animal researchers have managed to find evidence of "choosing sides" among lower primates in the wild. Wild mountain gorillas in Rwanda have been observed showing a specieswide predilection for their right arms, much like humans, and a study of the bones of wild rhesus monkeys found that their right arms were slightly larger than those on the left in areas where important muscles were attached—a sign of disproportionate use. Even brain imaging provided some proof, with scans of lower primates revealing that the left hemisphere's motor cortex is larger and more developed, offering more evidence of dominance and an evolving brain. [17]

All of this has been dismissed outright by those who say animal handedness is simply incomparable to human handedness. This is the position of some primatologists and linguists, including Noam Chomsky, who have fundamental problems with the idea of primates "choosing sides." Given that a collective handedness tilt is a sign that the brain has specialized in one hemisphere, these early primates would have been wired for the development of language, and for these scientists it is impossible to reconcile the capability of language with the reality that these primates don't actually "speak."

Despite this skepticism, other scientists maintain that our closest cousins share our neurological underpinnings. In the late 1990s, New York neurobiologist Patrick Gannon conducted brain-imaging studies on chimps and found that a certain section of their brains, the part used to create and comprehend language in humans, was bigger on the left side than on the right, as it is in human brains. This suggests a special function. "Chimpanzees possess the anatomical neural substrate for 'language' . . . essentially identical to that of humans," Gannon concluded. Furthermore, structural brain-imaging studies of great apes have found that the area that would be responsible for language is larger in the left hemisphere of their brains than it is in the right. This offers evidence of asymmetry and a sign that a language headquarters may be "growing" amidst the tangled wires. Duane Rumbaugh and Sue Savage-Rumbaugh have built on this research at the Language Learning Center in Atlanta, where they've found that chimps can understand the different meanings of sentences that vary only in their syntax, such as "put the raisins in the water" and "put the water in the raisins." [18]

While this appears to offer proof of an evolved language center, the ability to understand language is very different from the ability to produce it. Most scholars, including MacNeilage, concede that chimps barely get past the protolanguage stage, the level of a one- or two-year-old child. They can produce thirty-six different sounds, each with a different meaning, but most scientists believe they cannot produce a sequence of these sounds to create a meaningful "sentence."

Sue Savage-Rumbaugh is one of the exceptions. She has taught apes to "speak"' using a pictorial keypad that transmits words through an electronic voice synthesizer. The smartest of her bonobo chimps have a vocabulary of 250 words and, according to Savage-Rumbaugh, have constructed relatively complex sentences. In the end, however, the physiological reality is that chimps simply lack the specially constructed throat machinery, or voice box, that makes the intricacies of human speech possible. [19]

In addition to the vocal limitations of the chimpanzee, our relative brain developments may have followed different evolutionary trajectories when the chimp and human went their separate ways, causing the human brain to become three to four times larger than the chimp brain. Along the way, we developed advanced capabilities such as contingency planning, logic, and structured thought, all activities that require the ability to sequence, which, as William Calvin pointed out, is typically a function of the left hemisphere. [20]

Leaping, Posture, and Grammar

Almost a decade before Bill Hopkins started to document the right-handed leanings in his chimpanzees in Georgia, another animal scientist in Tennessee was noticing something else. It was 1980, and a man named J. M. Warren had just published a summary report of his research with rhesus monkeys, which concluded they showed nothing approaching human handedness. Only Homo sapiens, he concluded, could claim this higher trait. [21]

While Warren was receiving praise for his report and its conclusions, Jeannette Ward was in her primate lab at the University of Memphis studying visual perception in bush babies, prosimians that predated chimps. She was trying to determine whether the tiny raccoon-eyed Muppet Babies of the tree kingdom could tell the difference between a form and its mirror image. There was a popular theory at the time that an organism had to have a lateralized brain in order to distinguish reality from the looking glass. A human infant's brain becomes lateralized slowly over time, which is why young children often have trouble with mirror images of letters of the alphabet. But Ward's subjects came before monkeys, and conventional wisdom held that nonhuman primates did not have any laterality. "Imagine our surprise when they could tell the difference between the image and the mirror," Ward recalls. [22]

If they showed signs of lateralized visual perception, perhaps they were also hiding a division of labor for specialized motor functions. Ward began to test the bush babies to see how they reached for prey, and it didn't take long for a pattern to emerge: the prosimians she studied consistently used their left hands to capture moving insects and minnows—even if they preferred their right hand for other types of reaching. These findings were dismissed at first, but she continued to test other prosimian species, with the same results.

Ward was puzzled at first; how did her lab find such a strong hand bias when others had not? After reviewing all the previous literature, it came to her. Ward had used an existing piece of apparatus to test hand bias in the bush babies, a Plexiglas panel with openings placed high and low. In all previous research, there was no high reaching; the primates had been sitting or on all fours. Ward and her colleagues then tested the prosimians both ways and discovered that standing upright—or holding an unstable posture—increased the bias towards one hand. [23]

While Ward was observing the actions of their hands, Peter MacNeilage was studying their feet. Another pattern was emerging from his findings: when the prosimians leapt from branch to branch, the right foot consistently led the body in the majority of the animals. They would shift their weight to the left side in preparation for takeoff, balance themselves, and push off from the left leg base. [24]

MacNeilage went to visit Ward in her lab. He was developing a new theory about the origins of handedness, and her left-handed bush baby findings were integral to his right-footed leaping documentation. According to his new theory, choosing sides began long before the apes stood up. It began with the prosimian clingers and leapers who lived in the trees as early as 60 million years ago. One side had to lead out in action, initiating the movement from branch to branch. The other became more receptive, supporting, catching, holding things. The leaping side—most likely by chance—became the right. "The two sides of the body always do complementary things; one supports and the other acts," explains Ward. "Think of ballet dancers spinning on one toe. It's all about optimal coordination of body movement patterns."

With the right side of the body leaping and clinging to the next branch, the left hemisphere was gradually becoming developed to assume control of routine movement and the execution of motor tasks. Later on, as higher primates—monkeys and apes—dropped to the ground and began walking on all fours, the patterns continued when they squatted or stood upright and freed their hands. The animals were more likely to use their right hand to lead the action, manipulating branches, cracking open nuts, and holding fruit near the mouth—fine-motor tasks and practiced acts. The left hand would serve as part of the support system, creating a tripod with the legs. The more upright the task, the stronger the bias toward the right hand, as Hopkins would later discover in his lab with chimps and monkeys. [25]

This is the essence of the postural origins theory of handedness, first introduced by MacNeilage and two of his colleagues at the University of Texas. According to this theory, the root of laterality in humans and the reason it developed so strongly in us was to enhance the stability of postures and locomotion. Vestiges of the early posturing tendencies of the prosimians still remain today. "The link is actually not handedness but footedness," explains MacNeilage. "We have overwhelming evidence now that the left hemisphere's specialization for language is related more to a person's foot preference than to his hand preference." While footedness and handedness are also linked, it's worth noting that more than ninety percent of humans—including left-handed right-footers—control both language and body posture via the brain's left hemisphere. [26] (In fact, this author, a left-hander who is right-footed, discovered she was left-hemisphere dominant for basic language after undergoing the fMRI experiment, as mentioned in Chapter 4.)

While the posture theory explains how the right hand and arm became dominant when prosimians were clinging and leaping, it didn't seem to account for how prosimians such as Ward's bush babies developed superior left-handed skills for certain tasks—such as snapping up scattering insects and small animals. Still, a question remained as to how some primates could be mostly right-handed and others predominantly left-handed.

The left-handed snapping-up mechanism was actually the other half of MacNeilage's postural origins theory. While the early prosimians were reaching and clinging to branches with their stronger right hand, their left hand was free to shoot about grabbing at moving objects—swinging fruits, flying insects, tree gum, small vertebrates. MacNeilage called it the "smash and grab" reflex, lightning-quick reflexive arm movements needed to aim for and catch a moving object, much like the way a modern-day squirrel monkey scoops up a goldfish. Unlike the planned and controlled manipulation tasks of the right hand, such as turning a knob to get to food reaching for fruit and bringing it to the mouth, the body has no control over the trajectory of a ballistic smash and grab after it's initiated. For example, if an insect moves after a smash-grab movement has begun, the hand's trajectory or path of movement cannot be adjusted midaction (just as the trajectory of a bullet cannot be changed after it is released from a gun). The left hand must rely on the right hemisphere's visual-spatial abilities to anticipate the target's movement in advance, and aim accordingly. [27]

The differentiation of the roles of the two upper limbs came about when the prosimians started walking upright. Converting this new and unstable two-legged support system into an integrated single support system required alterations in the basic organization of the nervous system, which introduced the need for a more acute sense of balance to accommodate the motor instability of standing and walking on two legs. "Walking is basically falling forward," Ward explains. "We push off from one side but lead off from the other. If we raise our arm to throw something, the legs and trunk of the body must adjust slightly prior to the movement or else we fall over, our own weight unbalanced from the support center." Therefore the division of labor between the two sides in the human nervous system would have evolved from the bottom up, quite literally, when nonhuman primates started to assume upright posture. Ward concludes that this "postural stimulation" effect has continued to be a factor throughout human evolution, pointing to the right-handed bias still present in modern-day nonhuman primates, including the great apes. [28]

Today, the Galagos prosimians still leap across the treetops of the African rainforests, leading with their right foot and pushing off with their left, the same way the majority of modern humans, including at least half of all lefties, leap across a puddle. (Remember that more than half of all lefties have some form of executive-function dominance in their left hemispheres, too.) The human bias toward one side tends to increase when we are standing up. (There are very few people who can throw a strike with both hands, for example.) Similarly, at least a third of all left-handed writers, including President Gerald Ford, become more right-handed when they have to perform single-handed skills standing up, such as throwing a ball. [29]

While it was clear that evolution wired the brain for handedness and sidedness before language, there was still doubt about how motor skills as big as leaping, clinging, and throwing became so tightly connected to the finer movements of speech. The link lies in the relationship between the precise, repeated motions of the hands and the precise, repeated motions of the mouth and throat. "As the coordinator of clinging and leaping, the left hemisphere was originally specialized for general-purpose control of the body under routine circumstances," explains MacNeilage, adding that the right-hand capabilities and the left-hemisphere control of speech arose as separate offshoots of this general-purpose role. The first "words" would have come out of these routine movements—low-key sounds produced with each movement required to jump from tree to tree. Speech itself developed in a similar pattern, evolving with the sequential opening and closing of the mouth and the alternating of consonants and vowels and syllables. (Eating also played a role. "I think the syllable is a modification of the chewing cycle," adds MacNeilage.) Parallels have been drawn between the brain functions required to coordinate the sequence of a leap from tree to tree and the sequential ability that underlies language and logic in humans. An animal that remembered a successful series of movements—a diagonal jump from tree one to tree two—was more likely to survive and produce offspring. [30]

At the same time as the left hemisphere was coordinating the muscles of our posture and the movements of leaping and clinging, it was also becoming specialized for that prelude to language, gesturing, which also involves whole-body movements and shifting postures or poses. Even more intriguing was that these routine sequencing behaviors directed by the left hemisphere would be performed only under nonthreatening conditions. MacNeilage compares it to knitting, a "sequential output" that is not likely to be undertaken when predators are lurking. This is because the right hemisphere was evolving to become the emergency reaction hemisphere—the one that became activated when we had to respond to predators or bad news. This emergency response mechanism would preempt the left hemisphere's routine mechanism, making it difficult to perform the clinging and leaping and gesturing actions when perceived danger lurked. Traces of this right-hemisphere crisis-response mechanism can be found in modern-day humans. In neuroimaging and PET studies conducted by psychologist Richard Davidson and his colleagues at the University of Wisconsin, when subjects view photos of anxiety-inducing images such as spiders or an angry face, their right hemisphere's prefrontal cortex becomes most active, in a region important for hypervigilance under stress. [31]

Chicks, Frogs, and the Left-Darting Fish

While lower primates may be our closest link in the evolutionary chain, more remote species may have something to tell us about how we evolved to choose sides, even when it appears that there are no sides to choose. A quick slide down the food chain and a variety of evidence surfaces that all sorts of paws, claws, wings, and even fins receive preferential treatment. While most of these simpler creatures seem to have fifty-fifty odds of going left or right, a few species, like humans, share a specieswide preference for the right forepaw.

Side preferences have been documented in just about anything with appendages—from footedness in birds and toads to clawedness in lobsters and fiddler crabs. Fiddler crabs have been known to use their meatier claw for courtship rituals; in fact, to lure in the easily impressed females, male fiddler crabs wave their crusher claw up and down. "In other words they use them to display and perhaps signal their size and activity, not unlike us," explains C. K. Govind of the University of Toronto. Common toads have been observed consistently using their right front limbs to swat away annoyances near their heads—paper stuck in their nose, balloons tied to their heads. Picture the toad, if you will. This is not a creature that can throw a strike or rummage for earwax. Needless to say, this finding did not summon a phalanx of searches for a common ancestor of man and toad. As a dismissive group of scientists pointed out, toads have a few ugly belching habits linked to their uneven insides. Their stomachs are larger on one side than they are on the other, and occasionally they eject the contents of these lopsided bags right out the right side of their wide mouths—the easier to be cleaned with the slimy right forepaw. That would indicate that any paw preference in the toad is linked to their "visceral architecture," and not to any great divide in their brain. [32]

Chicks, fish, and other animals may not have limbs or even specialized wings that they could use as limbs, but they have eyes that serve different, independent functions—revealing different roles of the two hemispheres of the brain. Take the red rooster: he struts his way around the barnyard, swinging his wattle and angling his head this way and that to take in the changing farmscape. Beneath his clawed feet, a bed of straw, grass, and small stones often mix together with pieces of grain. When the rooster tilts his head to the side, comb hat flopping over, he trains his right eye to the ground, where it sorts through the blur of pale colors before instructing the beak where to go to snap up the grain. If he hears a caw sound in the distance, the left eye is aimed at the sky, scanning it long enough to ensure that a hawk is not within swooping distance. Once assured, the rooster can move along the chicken coop, until he comes across a sultry hen, and still stealing glances at the sky, catches a glimpse of her through his left field of vision. Suddenly excited, he plumps up his plumage, ruffles his saddle feathers, and jumps her. Luckily, for her offspring at least, she stumbled into his left field of vision, which transmits to the right brain. If she had entered the cocky cock's right field of vision, it is likely he would've shrugged his hackle feathers and continued his search for a snack among the pebbles. [33]

Different specialties of the eyes and their visual fields appear across many species, leading biologists to the dual attention theory of laterality, which is the idea that the two eyes and two cerebral hemispheres can specialize to perform competing functions, such as feeding and guarding, at the same time. The most marked examples of this division of labor can be found in chicks and fish. With eyes planted on the sides of their heads, they make for the ultimate "split brain" patients; each hemisphere processes a different view of the world, with the left eye sending information exclusively to the right hemisphere and vice versa. In humans, information about the world enters both eyes with a great deal of overlap, and each eye transmits information to both hemispheres. Everything in the field of view to the right of the nose is processed by the left hemisphere, and vice versa. To ensure that the brain doesn't get extraneous information, the fibers from the retina sort themselves out to separate right hemifield from left hemifield. Some fibers cross over whereas others, already positioned to see the opposite side of the world, do not cross. Unlike humans, chicks and fish have optic pathways that are almost entirely crossed, as each eye possesses its own distinct view of the world without overlap from the other eye. This allows scientists to draw conclusions about the nature of hemispheric arousal, activation, and specialization in these animals based solely on which eye the animal uses for any given task. [34]

In the late 1980s, a group of researchers led by Lesley Rogers of the University of New England, Australia, decided to experiment on baby chicks to see exactly how they process the world. Images of hens, strange lights, and chick-eating birds were alternately placed at a distance directly in front of the chicks. As they moved around the objects, the researchers recorded which way they turned their heads—and therefore which eye they relied on—to fixate on the object. If the object was a hen, the chicks would invariably turn their heads and use their right eye to stare at it. If the object was something new and different, such as a small light, they were more likely to use their left eye to examine it. Now, if the object was a hawk, there was no question which direction they would dart—an inner crossing-guard voice seemed to direct them: dart to the right, and keep your left eye on that marauder at all times. [35]

More recent studies of adult hens have found similarly predictable patterns. Forced to respond to a recording of a flying predator's alarm call, hens are more likely to turn their heads so that their left eye can look up at it. The right eye keeps its focus on the ground, staring at the pebbles and the grains in order to help the pecking beak discriminate between the edible and the inedible. Priorities may shift—the threatening sound of a swooping predator's caw is enough to shut down the appetite and any routine behaviors—but the division of labor allows the right eye to sift out food on a regular basis without having to worry about keeping watch. [36]

This left-eye orientation extends into the realm of mating, supplying matchmaking cues to the brain. Chicks treated with testosterone are more likely to copulate when viewing an attractive specimen exclusively with their left eye, which explains the red rooster's impulsively eager mating behavior. [37]

A similar type of phenomenon can be seen in our ill-mannered friend, the toad. In a study of visual perception, Italian psychologist Giorgio Vallortigara made an interesting discovery while watching toads tongue-strike their prey: a target had to enter the toad's right visual field before it would provoke a reaction—the snatching tongue. The toads were more likely to attack prey to their right side and ignore it on their left side, as if their left eye's visual field simply did not register the prey's presence. (Woe betide the fly whose flight path carries him past the right eye of a toad.)

This split role of perception reaches as wide as vertebrates without any limbs. Fish also have a special eye–and corresponding receiving center–with which they fixate on predators. Some vertebrates rely on the left eye to stare down their enemy; others rely on the right. A good example of this is the male mosquito fish. When faced with an obstacle like a clear glass bar barrier that has a group of female fish behind it, the mosquito fish will circle around the barrier leftward—so as to monitor the strange and potentially hostile barrier object with its right eye. Other species of fish, including poeciliid, also swim to the left of a potential predator. "Despite substantial differences between species in the general structures of the brain and visual pathways, particular functional specializations appear to be conserved throughout a wide evolutionary spectrum," sums up Vallortigara. This could be a spectrum wide enough to include humans. Cognitive psychology studies have found that people react more promptly to dangerous stimuli coming from the left side of each eye, which projects to the right hemisphere's visual scanner. An image of a famous person's face, for example, when presented in the left visual field, is more accurately and quickly processed than the same facial information presented in the right visual field, according to a study conducted by Jerre Levy and colleagues at the University of Chicago. These results reinforce the theory that the right hemisphere evolved in humans to become the stress-managing "hypervigilant" headquarters. This dividing of perceptual responsibilities not only makes the brain more efficient, but it can also improve the speed and skill of a "flight" response, according to Vallortigara. Pulling off a quick, slippery escape might require asymmetry of the muscles and nerves, which in turn could have produced other asymmetries at the sensory and perceptual levels. Of course, predators can appear on either side of their prey, but those who are specialized with a left-dart or a right-dart reflex can avoid indecision and therefore be physically more capable of quicker dodges, or perhaps even more artful fake-outs. [38]

Sharing the same eye and brain lateralization with your species offers evolutionary perks as well as drawbacks. When a shoal of bluefish sees an oncoming shark, they instantly and collectively make a sharp right turn, swimming in unison away from the sharp-toothed predator. But the bluefish who happened to inherit a left-jerk shift swims alone in the opposite direction. Who's more likely to get eaten? Interestingly, those "social" vertebrates that swim in schools (or drive on roads or march in formations) are more likely to share the same directional asymmetries across their species than are solitary operators, according to Vallortigara. Fish species divide between those that school and those that go it alone. The schoolers tend to share the same directional perceptions and reactions, whereas the rugged-individualist species have unpredictable lateralization patterns. In 1999, Vallortigara and his colleagues studied "detour" tasks in twenty species of fish and found that all of the most social shoaling species showed lateralization at the population level, while among the isolationists, only forty percent demonstrated a groupwide bias in preferred direction and processing. The rest showed only individual lateralization. It makes sense that the demands associated with living in groups, from courtship behavior to schooling, make it easier to share the same preferences. As for the drawbacks to sharing asymmetries, well, widespread behavior also makes a "shoaler's" actions more predictable, leading to a greater vulnerability. If the bluefish repeatedly dart to the right when they see an oncoming shark, over time, sharks learn to cut them off at the pass. Hence the fish with the left-dart reflex may actually be saved by going it alone to the left—call it the fake-out advantage. While the shark may also learn to anticipate these deviant left-darters over time, they will still benefit from attacking the masses, the proverbial "fish in a barrel." [39]

Humanity's Balancing Act

Individual humans also benefit—and suffer—from sharing the same directional bias as the majority in their species. At some point during human evolution (exactly when is a subject of debate between scientists), more and more people developed stronger right sides and a corresponding dominant left hemisphere. This collective tilt of the species toward right-handedness, driven by the infamous right-shift gene, presented a risk, according to Marian Annett's theory. As the pathways for motor controls, sequencing, and later speech were forming in the left hemisphere in the majority of hominids, the right hemisphere was becoming neglected. Yet we still needed the right hemisphere. With its more diffuse concentration of neurons, it was better at big-picture visual and spatial processing, including three-dimensional depth perception (knowing what lies on the other side of a cliff), and the ascertaining relationship of objects in space (mapping and navigating). [40]

To ensure that the right hemisphere did not fall into complete neglect, a gene had to be kept in the pool that prevented the species from tilting too far toward the right-handed/left-hemisphere end of the continuum. While the right shift (R+) gene was important for driving left-hemisphere dominance and right-handedness, two copies of the gene could lead to double the dose of left-hemisphere specialization—at the risk of a weak right hemisphere and left hand. Even righties need a left hand that is at least capable of helping out with important two-handed activities—steadying a stone during carving, swinging a branch to knock down fruit, or clocking a ball clear over the walls of Yankee Stadium.

To keep the species on an even keel, evolution worked it out so that the majority of humans would be balanced, receiving one right-shift gene (R+) and one counterbalancing "chance" gene that is less inclined to nudge the brain or the body in any direction (R-). This notion is referred to as human balanced polymorphism theory, and it is the evolutionary argument behind Marian Annett's genetic model. Evolution ensures there are many (poly) forms (morphisms) of a gene or trait so that a species does not become too lopsided in any one area. Those who inherit both forms of the handedness gene, R+ and R-, get just enough of a genetic directive to prune the right hemisphere down while building a left-hemisphere headquarters—dampening left-handed motor skills without squashing them completely. When it comes to their cognitive abilities, Annett argues that this group has the advantage of streamlining the language abilities in the left hemisphere, yet simultaneously maintaining the right hemisphere's specialties. [41]

On either side of this balanced group are the counterweights—the strong right-handers on one end of the scale and the strong left-handers on the other. The strong righties, with their double dose of brain-nudging genes (R+ R+), have overly sheared right hemispheres, which can lead to weak nonverbal skills, according to this theory. The lefties (R+- or R--) offset them on the other end of the handedness spectrum, as they lack any built-in mechanism to trim down the right half of the brain. In their case, the nonverbal brain is free to develop the capacities that the more biased members of the species lack. The risk is that R- will leave the lefties lacking in the verbal department.

While there are aspects of the counterweight theory that seem to ring true, it has many detractors. Critics have challenged its assumption that those on either end of the handedness scale have any inherent cognitive deficits—or even advantages—and point to multiple studies that fail to support the claims. And if a significant subset of left-handers (not to mention some right-handers carrying the same genes) has a more "randomized" brain, those individuals are likely to end up with some language faculties taking hold in the right hemisphere, and other typically right-hemisphere specialties ending up in the left half.

According to human balanced polymorphism theory, the role of the left-handed gene, R-, is to keep the right hemisphere's nonverbal specialties thriving. If those who inherit R- can end up with verbal capacities in their right hemisphere and nonverbal skills in the left hemisphere, then there is no guarantee that visual-spatial and other typically right-hemisphere abilities will become more developed. In other words, the more "randomized" left-handers cannot be counted on to counterbalance the species. Yet they are still with us, and this must be for a reason. [42]

While there is still debate about exactly how the human brain came to be asymmetrical, there is evidence that our evolution was driven by a combination of forces—the need for balance and efficient movement in an upright posture, the advantages of an accurate throw, the importance of sharp tools, and the benefits of gesturing and communication.

Yet, though laterality plays a role in all species, it also seems clear that most animals are partial to the right, and that relying on the left side is a relative rarity. But if, as many people continue to think, the forces of evolution work against negative traits to drive us all toward one uniform design, why did a significant minority of our population remain "backward"?

7

The Triumph
of the Lefties:
How Southpaws
Survived and Thrived

*Whereas right-handers, because their genetic repertoire is more
predictable and somewhat more limited, are necessary for our
species . . . they may not provide the variation in behaviors
or ideas, potentially, that you could argue left-handers might.*

— Dr. Jordan Grafman, National Institutes
of Health, on NPR, March 2002

Isabella d'Este, the mysterious woman who was the subject of
the *Mona Lisa*, was described by those who knew her as having
a "strange and complicated personality." According to da Vinci
biographer Charles Nicholl, the wealthy marchioness could be

imperious and petulant, ridiculous and yet sharply intelligent. Over the centuries, Da Vinci's ability to capture the contradictory nature of his subject has been analyzed by historians, artists, photographers, psychologists—and neuroscientists. In spite of his many inventions and scientific discoveries, he is perhaps most famous for his exceptional ability to read and translate the subtleties of Mona Lisa's complex expression, an ability that has been attributed to the fact that he was left-handed. [1]

Freezing a face in time, or turning letters inside out to create mirror script, may not seem crucial today. But the Renaissance man's unusual talents, as well as those of many other left-handers throughout history, shed light on how and why the trait survived, and why it continues to thrive.

Science has offered countless theories as to when and why the human race became predominantly right-handed. A century's worth of studies set out to determine the pathologies of the aberrant southpaws, to understand the limitations and flaws that leave them weaker than the right-handers of the world. Yet these theories and queries overlook the vital question that we've been asking all along: if evolution weeds out the weak in favor of the fittest, how have these dismissed and disadvantaged left-handers survived?

While some theories suggest that the left-handed and their genes are here simply to serve as a counterweight to the species, evidence is everywhere that the left-handed have played a far more dominant role in the evolution of mankind.

Whether they have standardly organized brains or creatively shuffled arrangements, strong left-handedness or mix-and-match tendencies, those whose genes go against the grain of right-handedness have made important contributions to many of the great leaps of humankind. Now faced with the modern world, their unusual and unpredictable predispositions may be more essential than ever. Threading together a combination of theories and evidence, from the early savannas to the baseball diamonds,

The Triumph of the Lefties

from Renaissance Italy to the Revolutionary War to the White House, a story emerges of how left-handedness has played its own part in advancing civilization. It may even explain why the trait is on the rise.

Hunters, Sharpshooters, and Babe Ruth

In the Old Testament, it is written that Benjamin had a tribe that consisted of seven hundred "chosen men left-handed," every one of whom "could sling stones at an hair breadth and not miss" (Judges 20:16). Benjamin might have been the biblical forefather of the Yankees' coach, Joe Torre. Long before ace pitchers were recruited to bring home the World Series Championship, slinging stones with the left hand helped early man bring home a bigger supper.

As you might recall, the evolution of throwing was believed to have played a vital role in the advancement of the human species; the demand for precision and accuracy in throwing led to a high level of specialization of the motor apparatus in the brain, which paved the way for the finer coordination of the motor apparatus for speech. The need to issue large numbers of motor commands to the muscles, and to do this rapidly and precisely in both speech and hand control, has led researchers to link language and handedness to left-hemisphere specialization—and the corresponding right-handedness. Yet somehow an entire biblical tribe, not to mention eleven to fifteen percent of modern-day humans, ended up with their motor headquarters in the right side of the brain, and their throwing arm on the left side of the body. "One of the highest achievements of the brain is the act of throwing," says Michael Peters, professor of psychology at University of Guelph in Canada. "It totally amazes me what pitchers can do, how close the ball consistently comes to where it's supposed to be," says Peters. "You only get that kind of accuracy if you have a great pile of neurons firing together, pooling their outputs to yield accurate timing for the act of releasing. To do that, you've got to have a very specialized motor system." [2]

Throwing accurately requires that complex motor system, involving a motor-planning director and motor-output centers in *both* hemispheres of the brain. Regardless of which hand is used to throw a ball, both hemispheres must cooperate to allow the body to aim and hit a target—especially a moving target. It's a highly developed skill that requires repeated practice, which is why very few people can throw well with both hands.

The early ancestor who happened to throw his first rock with his left hand has his modern equivalent in baseball legends Babe Ruth, Ted Williams, and Sandy Koufax. More recently, Andy Pettitte, David Wells, and Randy Johnson are examples of left-handed "chosen men," with piles of neurons firing together from the right hemisphere of their brain. Their ancestors were very likely among the overlords of the savanna, accumulating prized dead birds the way these celebrated southpaws have collected MVP trophies.

Throughout Major League Baseball, left-handed pitchers show up at nearly three times their rates in the general population, representing more than one-third of all pitchers. If an All-Star Game lined up lefties against righties, many believe the northpaws wouldn't have a shot. "Give me all the left-handed pitchers and I'll beat you to death," Clark Griffith, the first manager of the New York Yankees, told Gene Mauch, manager of four different major-league teams in twenty-six years. After the Yankees' momentous loss to the Red Sox in 2004, several sportswriters told George Steinbrenner exactly what he needed to do in the off-season: snap up a left-handed pitcher. Steinbrenner took their advice, snapping up the biggest lefty of them all, Randy Johnson, winner of four consecutive Cy Young Awards. (The Red Sox followed with their own left-handed catch, pitcher David Wells.) [3]

Many sports analysts and a few evolutionary psychologists have argued that the left-handers' pitching prowess is strictly a numbers game.

The average batter is more accustomed to a right-handed throw. This leads them to be less prepared for the mirror-image angle and gives them more trouble at the plate. The throwing arm of the lefty pitcher is also hidden from the right-handed batter's view, which makes it harder for the majority of sluggers—and prehistoric prey—to gauge the pitch that's about to be fired their way.

Other experts—and a few catchers—believe the left-handed pitch is mechanically different. There is the geometric difference. As sportswriter Gerry Fraley of the *Dallas Morning News* explains, the angle from which a lefty throws makes it more likely that his pitch will cut inside the plate for a righty hitter—a tougher zone from which to hit. Conversely, the left-hander's breaking pitches (the curves, sliders, and cutters) swerve away from the left-handed hitter, out of his power zone, while a good lefty fastball arcs back in to the inside of the plate and jams a left-handed hitter.

Ultimately, the left-handed pitcher's strengths may reach beyond the physical manifestation of the pitch, with skills that are far more intrinsic, rooted in the unusual arrangement of their brains. Since more than half of all lefties develop the standard "right-handed" brain organization, their visual-spatial headquarters, including three-dimensional perception and spatial planning, develops in their right hemisphere, adjacent to their dominant motor headquarters. This arrangement—the close proximity of visual-spatial processing to motor skills—can help facilitate the orchestration of the precarious pirouette that is the wind-up-and-hurl. [4]

Growing up surrounded by right-handed equipment, instruments, appliances, and tools, lefties give their nondominant side more exercise than the average righty. Biomechanics research has revealed that training the nondominant side of the body actually enhances the dominant side—something known as the cross-training effect—since the body's neural network is integrated on both sides. Rafael Escamilla, director of the Human Performance Laboratory at Duke University Medical Center, explains that in sports movements such as those a baseball pitcher makes, the whole body contributes to the motion, not just the dominant side. "You want symmetry in your body and you don't want to neglect a particular area." Greater symmetry improves balance and coordination, which improves precision and accuracy. The left-handed, who spend a lifetime training their weaker half, are cross-trained before they even step up to the pitcher's mound. [5]

The right-handers, with their less-developed right hemispheres and underused left sides, are less likely to match the strength and balance of the left-hander, at least in absolute terms. According to Marian Annett's theory, the role of the right-handed (R+) gene is to limit growth of the right hemisphere, which means there is some loss of brain power on this side in right-handers that limits the abilities of their left sides. "Hence there is likely to be physical imbalance, relative weakness of the left arm and hand, left leg for kicking etc., and perhaps relative weakness of right hemisphere-based visuospatial skills," says Annett. [6]

Still, it is the relative strength of the left-hander's more developed right hemisphere, particularly when it comes to those standard visual-spatial skills, that may be the true secret to their pitching prowess—and their early survival on the savannas. Most people, left- or right-handed, rely on the left cerebral hemisphere to organize and plan controlled movements, turning on and turning off different muscle groups while coordinating the timing and intensity of muscular forces. As Digby Elliott, professor of kinesiology (the study of movement) at McMaster University in Canada, explains, "This ability is very important, not only for very precise finger movements, but also for production of precise oral movements." In addition to using the left hemisphere for these processes, the left-handed receive an additional benefit from the dominant right hemisphere. According to Elliott, the right hemisphere plays a role very early in the movement organization process, locating the spatial coordinates—the strike zone—and making split decisions about where and when to release a ball to change course. So if a batter makes a sudden change in movement—crouching down to bunt—the right hemisphere can redirect the last-minute release.

Some believe that when the target is a left-handed batter, the southpaw pitcher will have an easier time sizing up his or her spatial coordinates—that is, aiming—as information that falls into the left half of the visual field is typically processed more quickly and more accurately by the right hemisphere. As headquarters of spatial planning, the left visual field/right hemisphere connection can be useful when it comes to locating the spatial position of a target. Aiming can also play to the strengths of left-eye dominance, a trait claimed by approximately seventy-four percent of left-handers. Information received by the dominant eye is processed faster than that received by the nondominant eye—even if the dominant eye is visually poorer. A dominant left eye relays visual cues directly to the right hemisphere's spatial planning headquarters, and when it comes to converting the cues into action, the right cortex has a "qualitatively distinct" advantage for the translation of limb segments through extra personal space, says Elliot, and this is expressed as a left-hand advantage for speed or accuracy measures. [7]

In 2002, two Turkish physiologists set out to study eye and hand dominance in handball, a sport that requires alternate use of the left and right hands for "swatting" the ball, as well as alternate use of the left eye (which must lead the aiming when the body is turned to make a right-handed swing) and the right eye (which has a better view when the body is turned to the left in preparation for a swing). Their findings, backed up by a separate study in 2003, suggested "a left-handed advantage over right-handed in the focusing of left eye and the left eye–left hand visual reaction times, but do not show a right-handed advantage in the focusing of right eye and the right eye-right hand visual reaction times." The researchers attributed the left-handed edge to a better performance in right-hemisphere visuomotor activities. [8]

While there's no conclusive evidence that left-handed pitchers generally have an innate edge over right-handers, it appears that certain subsets of lefties come with a distinct, built-in pitching potential. Of course, how this potential is realized depends on many factors—from those random variables that shape our early development to the presence of left-handed mitts and coaches. But what about behind home plate?

Fishermen, Catchers, and Mike Piazza

It may seem like a grand leap, from squirrel monkey to Mets catcher Mike Piazza, but we are about to make it—leading with the right foot. Dating back to those early primates, the prosimian clingers and leapers who jumped around the treetops in the dense forests, the right arm was evolving to become the stronger of the two, the gripper that clung to the trees, leaving the left hand free to develop the swift reflexes needed to capture falling or scattering tree grubs. "It was quite crucial to survival," says Peter MacNeilage of the left-handed skill that is the smash-and-grab reflex.

It's a survival skill that still exists today, and according to some scientists, it even manifests itself in species that are considered to be right-handed. Primatologist Jeannette Ward's prosimian bush babies may prefer their right arms when they reach for stationary food, but they are faster at catching moving insects with their left hands. Squirrel monkeys consistently catch swimming minnows and other moving objects with their left hands while propping up their body weight with their right.

For these animals as well as humans, the left hand's strength at snapping up moving objects stems from the limb's connection to the right hemisphere, with its specialty for spatial tasks and synchronized timing. "You had to aim properly in space to manage to stay alive with ballistic reaching," explains MacNeilage, referring to the bulletlike movements of a hand shooting out to catch a shimmering fish or a snapping predator's tail. In addition to the speed of the reflex itself, successful movements depend on the ability to anticipate the direction a tiny swimmer will take, or the trajectory of a flying mosquito, requiring good mapping and aiming in three-dimensional space.

Right-handed humans also reap advantages from the mercurial left-handed grab. MacNeilage and colleague Win Lee demonstrated this when they put a group of right-handed students through a simple reaction-time exercise. From this, they found that the subjects were markedly faster (thirty milliseconds) snatching a moving target with their left hands than with their right. (The next time you see a right-hander haunted by a buzzing mosquito, suggest that he use his left hand to swat at it—he might just avoid grabbing at air.) [9]

Though our survival may no longer depend on the swatting, smashing, grabbing and scooping of moving insects and fish, evidence of the left hand's lightning reflex remains today—even among right-handers. "It's the left jab in boxing," suggests MacNeilage, who sees the quick left punch as a product of this smash-and-grab reflex. Another vestige of this reflex can be seen when a right-handed catcher like Mike Piazza reacts with the glove on his left hand, adjusting his mitt at the very last second to snap up a ball spiraling unpredictably off of a bat. It's worth noting that just about every major-league catcher is right-handed and therefore catches with his left hand, yet another example of the overlooked importance of the left hand. (Of course, since most batters are right-handed and stand to the left side of home plate, they interfere with a lefty catcher's ability to throw out base stealers at second, so this is another reason catchers are predominantly right-handed.)

One of the few lefties to play catcher was Babe Ruth. As a boy at St. Mary's Industrial School in Baltimore, Ruth's first position in baseball was behind home plate, but despite his substantial talent, even he was forced to use his left hand to catch. In his own words: "We had no catcher's mitt built for left-handers, of course. We were lucky to have any kind of mitt. I'd used the regular catcher's mitt on my left hand, received the throw from the pitcher, take off the glove, and throw it back to him left-handed. When I had to throw to a base, trying to catch a runner, I'd toss the glove away, grab the ball with my left hand, and heave it with everything I had." [10]

Navigators, Protectors, and Shrinks

Unlike some of their more highly evolved descendants, the early hunters and gatherers didn't neglect their left hand. They valued its smash-and-grab reflex, which brought in the minnows, grubs, and salmon. Of course, they still needed to find these salmon-rich rivers and bug-rich forests, and while they searched, they had to keep an eye out for potential predators hiding along the route. Before there were road signs and global positioning satellites, our survival depended on having a good mental map of our surroundings, gauging which direction we came from and the quickest route back to the cave. This was especially important among hunters and gatherers, who moved often. "If you're a mammal surrounded by horrible things, it's quite important that you have a safe burrow to go to, and that you use cues—rotten trees, mountains, and so on—to find your way home," explains Oxford's John Stein. [11]

It also helps to remember the salient details of such cues. As psychology professor Stephen Christman found when he studied episodic memory, the "mixed-handed," which includes the majority of lefties, are better at recalling the vivid details of events. "They were better able to explicitly remember the time and place at which a new memory was encoded," he explains. Christman attributes the advantage to a greater integration of information between the two hemispheres, and cites findings that the mixed-handed have a thicker bundle of fibers connecting the two sides of the brains. "A single hemisphere can remember an event, but not as quickly or as accurately as when you have both hemispheres working together," he says, adding that the left hemisphere appears better at remembering facts, while the right hemisphere adds in color and context.

Greater integration between the hemispheres also may keep a person from getting lost. Another recent study found that left-handers are significantly better than right-handers at remembering images that involve direction, such as which way the queen is facing on an English pound. Recalling which direction a nose—or a rotten tree branch—pointed could be the difference between returning safely to the protective tribe or wandering the wilderness alone. There is also contemporary evidence that apprehending and navigating through uncharted terrain may favor left-handed tendencies. In the 1970s, an American sociologist named John Dawson studied nomadic hunting and fishing peoples such as Inuits and Australian Aborigines and found that they displayed higher rates of left-handedness than the more rooted agricultural communities such as the Temne and the Chinese Hakka. [12]

Throughout human evolution, the ability to decode and map a landscape has been aided by greater use of the right hemisphere. It's a dichotomy most apparent in animals whose visual fields do not overlap—the birds and fish with eyes on opposite sides of the head. They depend on the brain's division of labor to avoid conflicting responses between the two hemispheres. As authors John Bradshaw and Lesley Rogers discuss in their book, *The Evolution of Lateral Asymmetries, Language, Tool Use, and Intellect,* the left eye/right hemisphere system in birds is specialized for processing topographical information, the kind needed to build a map of the environment and respond to changes in it. [13]

Similar capabilities were crucial to our ancestors. Early man came face-to-face with a wild assortment of creatures, predators the size of a Winnebago, ominous competitors. Keeping genes in the species that promoted the development and use of the right hemisphere allowed our ancestors to sort the hostile from the friendly, and react accordingly. Just as in chicks and fish and other species that rely on their left eyes to scope for predators, there is evidence that the left visual field in humans detects and responds to danger more quickly. Since the majority of left-handers are also left-eye dominant, their response mechanism would be quicker still, alerting the right hemisphere to a hostile presence in time to act. [14]

Face recognition and the processing of emotional cues usually falls under the right hemisphere's command, part of what's known as spatial integration. When researchers sever the corpus callosum in a rhesus monkey, only the left eye, controlled by the right hemisphere, can discriminate between photographs of different monkeys' faces— important when sorting mates from predators. Similar to the rhesus monkeys, human patients who've suffered brain damage in their right cortex tend to have trouble reading faces and interpreting other people's intentions or responding to contextual cues. Without full access to the lobe's processing mechanisms, the interpretation of complex visual displays can be blocked, as Robert Ornstein explains in *The Right Mind*. People become less able to discern the "minute changes in facial coordinates" that lead to an expression, and this can mean life or death. "Expressions provide important information about what an animal is going to do, about whether it wants to fight, to mate, whether it's indifferent, whether it's an animal of one's own species, whether it's harmful or harmless," says Ornstein. [15]

In prehistoric times, this more developed "facial coordinates processor" would have been helpful for sorting friend from foe, detecting the less than benign intentions of an approaching hominid— the subtle arch of an eyebrow or slight quiver of an upper lip. Today, the ability to read a person and decode his or her true intentions is the focus of research and training in the field of law enforcement. A well-known study led by Paul Ekman in 1991 found that judges, customs officials, police detectives, FBI agents, and forensic psychiatrists were no better than chance at judging the honesty of videotaped speakers. (Only members of the Secret Service performed significantly above what would be expected from guessing alone.) As a follow-up study, several Canadian scholars set out to test a general population in 2002 to see if they could isolate different traits that might contribute to better judgment when it came to lie detection. They found that left-handers were "substantially better at detecting deception than were right-handers." Although both handedness groups reported using a similar number of speech, body, facial, and "vague" cues, left-handers apparently relied more on visual cues and less on the content of a person's statements to decide his or her credibility. The researchers speculated that the left-handers' "flexibility in thinking" and use of a "visual-based process" could explain their greater sensitivity to the subtleties in communication. [16]

The emotional dimensions of language and expressions—such as the differences between a look of anger and one of fear—also tend to fall under the jurisdiction of the right hemisphere. Left-handers who inherit a more randomly organized brain may be wired in such a way that auditory processing and emotional reasoning end up adjacent to each other—an arrangement that would be useful for distinguishing between vocal intonations that communicate different emotions, hostility versus anguish or elation. This might explain why psychiatrists show up in studies as disproportionately left-handed. It might also justify how the left-handed da Vinci was able to capture the flickering facial coordinates and complex emotional expression of the *Mona Lisa*. [17]

Again, our prehistoric ancestors provide examples of how crucial a well-developed and well-integrated right hemisphere is to our livelihood. When the well-being of a tribe might be threatened by outsiders, the group needed to be warned when a guttural growl of anger or a face full of fury was detected. The early hominids who inherited the R+ right-handed gene might have been the first to turn grunts into distinguishable sounds, perhaps sounding the first

"Run for the hills!" Those who inherited the R- "chance" genes may have followed suit, or they may have developed better nonverbal communication styles, expressing themselves more with faces, gestures, and physical acts such as the mimicking of the predator and its state. Silent communication would have been more effective when predators lurked nearby. These are the kinds of skills that would make someone a good mime, like Marcel Marceau (a lefty), or a great physical actor, such as Charlie Chaplin and Harpo Marx (lefties). You can find their descendants in the more recent assortment of "whole body" actors, left-handers Dick Van Dyke, Carol Burnett, Tim Conway, and Richard Pryor, and more recently the rubbery-faced Jim Carrey, the hyperactive Ben Stiller, and *Seinfeld's* Michael Richards, the actor who created the apoplectic Kramer. Perhaps it's not surprising that studies of students enrolled in acting classes have found disproportionate numbers of left-handers. [18]

Ultimately, it isn't enough to physically communicate danger; the early hominids had to react to a threat before it reacted to them. Studies have found that people—presumably ones long-since dead as well as modern humans—are quicker to react to an angry face flashed to their left visual field. "The right hemisphere is designed to make quick judgments of other animals' expressions," says Ornstein, adding that the right hemisphere's role in overseeing the larger muscle movements of the arms and legs might explain its reputation for governing the "withdrawal" response. "Fear and anger may require running, fighting, or something big in terms of consequence or in terms of the limbs," says Ornstein. In research conducted by neuroscientist Roger Drake at the University of Southern California, the right half of the cortex has also been implicated in the role of risk perception. [19]

Given that the left-handed and the mixed-handed (this includes the self-classified right-handers who do a few things left-handed) have greater access to their right hemisphere, Christman and his colleagues set out to determine whether these groups might be more inclined to take risks. They conducted an experiment to determine the extent to which a person's perception of an activity's potential risks versus benefits more strongly predicted their likelihood to engage in that activity. "We found a robust effect of handedness: strong right-handers' likelihood to engage in activities was determined almost solely by the activities' perceived benefits: if it is potentially beneficial, they'll do it, regardless of the risks. On the other hand, mixed-handers' likelihood to engage in activities was determined almost entirely by the potential risks: if it is risky, then they won't do it, regardless of the potential benefits. Put another way: mixed-handers are more risk-averse than strong handers." [20]

A greater risk aversion may be tied to the emotional division of labor in the two hemispheres. We know from neuroimaging studies of humans that the brain's right hemisphere tends to specialize in "withdrawal" responses, activating when a person looks at images of angry faces or spiders. This response was adaptive, facilitating our ability to retreat from a potentially threatening situation. Of course, there are risks associated with a more developed withdrawal system. Richard Davidson's studies at the University of Wisconsin have found that children with greater relative right-sided activation in the prefrontal cortex show more inhibition and wariness in their behaviors. From an evolutionary perspective, greater inhibitions and apprehensiveness could hamper the ability to compete for food or hunt for prey. [21]

Perhaps in order to cope with its weighty evolutionary responsibilities—all of this predator monitoring and crisis preparedness—the right hemisphere also assumed responsibility for the subtleties of humor, including the multiple meanings of words and phrases or the unexpected turn of phrase. Through studies of brain-damaged patients, we learned that the right cortex is important for understanding sarcasm and jokes, and more recently, brain-imaging studies have found that the left-handed tend to have larger "humor centers," activating a broader region of the right hemisphere—and some of the left—when listening to jokes. [22]

For all those lefties who might be worried about their evolutionary capacity for overworrying, however, the good news lies in the corpus callosum—that bridge of cables between the two hemispheres that tends to be bigger in the non-right-handed. "Since the two hemispheres appear to operate in a more balanced and integrated fashion in mixed-handers, I suspect that they should exhibit more even-keeled moods, e.g. less extremes in emotional experiences, both positive and negative," offers Christman, who points to studies that have found that the left- and mixed-handed are actually less anxious than are strong right-handers. [23]

And for the lefties who still find themselves up late worrying about something—a swooping predator or that scowling boss—adaptation is just a remote control–click away. The late-night comedy circuit is home to a surplus of southpaws exhibiting their highly evolved humor, among them Jay Leno, Comedy Central's Jon Stewart, and *Saturday Night Live's* Tina Fey. (The in-between channels can be counted on to offer reruns courtesy of left-handed comedians Jerry Seinfeld and Drew Carey, or lefty Matt Groening's *The Simpsons*.)

Warriors, Fencers, and Rocky Balboa

In *Rocky*, when the defending champ first considers the entertainment value of taking on the washed-up "Italian Stallion," his manager tries to discourage him. "He's a lefty. I don't want you messin' with no left-hander. They do everything backward." Doing everything backward, as Rocky learned, can be a very good way to stay alive—on the savanna or in the ring. It was the battered Rocky's backward left hook that took the champ by surprise and dropped him to the mat.

Of course, doing everything backward can also leave a person vulnerable. As human societies shifted from the nomadic to the agrarian and tools became more sophisticated, the species began to benefit from the consistency of right-handed conventions and creations. Advancements in weaponry also lessened the need for the strike-zone aim and smash-and-grab reflexes that favored the left side, while skills relying on spatial integration became less essential as societies rooted themselves in the soil. With this more stationary form of civilization came the organization of armies, banding together as the Roman legionnaires did, marching into battle shoulder to shoulder. Armies favored synchronicity and conformity: sword in the right hand, shield in the left. If a soldier left a gap in the protective wall of shields (held in the right hand), he risked not only his own life but that of others behind him. [24]

But not all combat was organized. While against-the-grain instincts might have created problems for left-handers in the conformist "schools" of right-handed armies, their unusual side preference also provided them with a tactical advantage. In one-on-one combat, for example, weakness in numbers can be a strength. Left-handers, quite literally, come from left field, surprising and confounding their adversaries. "The key to a hockey fight is the first punch," Boston Bruins hockey star Wayne Cashman once told an interviewer. "When you're a lefty and they're looking for the right, it helps." This is what evolutionary theorists refer to as the infrequency advantage: the fewer the numbers, the greater the edge in the arena of combat, where the majority is less prepared for what the minority is going to throw at them. The same paradigm prevails in baseball—hitters are more comfortable receiving pitches from right-handers, since right-handed pitchers occur with greater frequency. [25]

Michel Raymond, a biologist at the University of Montpellier in France, put the theory to the test in 1995. The left-handed are at an advantage each time they fight a right-hander, he argued, as long as they remain in the minority. Teaming up with colleagues in Copenhagen and Lyon, Raymond studied data from two different categories of sporting contests: man versus the clock, and man versus man. For sports involving man against the clock—the solo sports of swimming, running, and track and field—they found no differences in performance between the left- and right-handers. But when sports involving face-to-face duels were sifted from the data, the story changed. Most striking were the numbers produced by lefties in fencing, a one-on-one combat sport. Half the men and one-third the women who reached the quarterfinals of the world championships of fencing between 1979 and 1993 were left-handed, according to the study. Jousting and sword fighting were used to settle disputes in ancient China, Medieval Europe, even as recently as Revolutionary America. From samurai warriors to swashbucklers to pirates, from Alexander the Great to Andrew Kerr, a skilled left-handed swordsman had a survival advantage over his right-handed counterpart—simply because of the importance of the element of surprise. [26]

More evidence of a left-handed warrior advantage arrived in June 2000, when a team of Greek scholars found a surplus of southpaws among university-level athletes in sports such as boxing, fencing, football, basketball, tennis, judo, karate, and volleyball. The closer and more confrontational the physical interaction of the opponents, the greater the numbers of left-handers. As with the French study, the Greeks did not find more lefties in the noncombat sports of cycle racing, running, diving, gymnastics, or skiing (sports that do not require single-handed skills), and this led them to conclude that the left-handed edge was strictly strategic. Similar to their advantage in baseball and rock throwing, a "backward" lefty benefits from the psychological and tactical perks of coming at opponents from unexpected directions and angles. "This requires right-handers to repeatedly reverse their usual strategies when facing a left-hander, but it also necessitates frequently fielding unfamiliar attacks," the study concluded. [27]

Apparently this advantage is less evident once athletes reach the professional level and right-handers become more accustomed to playing against left-handers. David W. Holtzen, Ph.D., a clinical psychologist and instructor at Harvard Medical School who has studied left-handed athletes, argues that the right-handed pros are very familiar with the left-handed playing style. "Because their livelihood depends on it, they have strategies against it," he says. And just as prehistoric warriors might have passed along campfire tips about how to duck the "curve rock," today's Wimbledon finalists and Olympic fencers, boxers, and judo masters have manuals and coaches and practiced techniques. [28]

Other experts believe there is more to the left-handed combat advantage than simply the art of surprise. French neuroscientist Guy Azemar analyzed handedness in "opposition sports" in a 1993 study for the National Institute of Sports and Physical Education in Paris, and came to the conclusion that the southpaw's "upper hand" was innate, particularly when it came to sports involving "fine and fast" movements. For example, fighting with a broad sword, like a samurai, would favor the right hand because swinging the broad sword is a power move. Fighting with a fencing foil that requires razor-sharp reflexes plays to the strengths of the left hand, since it taps the same "ballistic" skills studied by Jeannette Ward in early primates and Peter MacNeilage in latter-day humans. [29]

In addition to quick reaction time, a few other inherent abilities also factor into the winning warrior equation. In 1999, Holtzen teamed up with colleagues at Cambridge University to test the hypothesis that the left-handed share a knack for visual-spatial, whole-body motor, and attentional visuo-motor skills. They chose to measure left-handed prowess in tennis, a one-on-one "combat sport" that requires all three of these skills: the whole-body motor functions that include fancy footwork, positioning and striking a ball, directing attention to the ball as it is returned, and apprehending the width and depth of the court in time to respond accordingly. [30]

Holtzen and his colleagues found that rates of left-handers were two to five times higher than expected among the top-ranking professional tennis players over a thirty-two-year span—from 1968 to 1999—among the top ten, Grand Slam finalists, and "world number one" tennis players. Since years of playing against lefties did not seem to help the righties reduce the numbers of southpaws at the top, they concluded that the left-handed advantage must be intrinsic, rooted in their distinct neurological arrangements—in particular the subset of left-handers who have their three-dimensional processing and motor-planning skills situated in the right hemisphere.

As Holtzen reviewed the findings, he came to view tennis as a task of "kinetically applied geometry," one that taps the same hexagon-making skills used in tenth-grade math. (Interestingly, at least one study has found that left-handers score significantly higher on tests of geometry skills.) "I think of tennis as a visuo-spatial-motor activity, the use of one's body to carry out geometrical tasks, geometrical 'problems' solved by the right hemisphere," he explains. Visualizing the court in three-dimensional space, anticipating the trajectory of each shot, and preparing for response are all applications of geometry, and they all fall under the jurisdiction of the right hemisphere. [31]

Timing is also critical. Azemar, the French neuroscientist who also studied tennis, points out that left-handers are likely to have visual processing, movement control, and three-dimensional planning in the same hemisphere. "In a right-hander, visual information has to transfer from the left hemisphere to the right, adding an extra twenty or thirty milliseconds to the reaction time—which can be significant at the elite level of sporting competitions." [32]

Finally, just as adapting to a right-handed world provides rock hurlers and major-league pitchers with the advantage of a more balanced whole-body movement, it may also provide an even greater "net sum total" in skill level when it comes to two-handed endeavors, which these days includes tennis. This is certainly true in one-on-one showdowns with one-two punches—such as boxing, judo, and karate, where having a relatively stronger nondominant hand would give an additional edge. Several experiments have found that the right hand of lefties is quicker, stronger, and more skilled than the left hand of righties. [33] Another study by the French Institute even found that left-handers have faster eye-hand coordination with their nonpreferred hand than right-handers do, giving southpaws an absolute advantage over right-handers in motor speed. [34] Over time these skills build up, as the more frequently left-handers take on everyday tasks with their right hand, the better their nondominant hand performs in sports.

Conquerors, Campaigners, and Alexander the Great

The surprise left hook and left thrust may have saved the lefty in one-on-one battles, but outside of the sporting arena and the schoolyard, dueling has not been the most fashionable way of settling disputes for several centuries. With the possible exception of Mohammed Ali's daughter and fighter pilot John McCain's kids, few modern-day humans can claim they arrived on the earth as a result of their parents' or grandparents' combat skills. In other words, if lefties survived only as a result of surprise left hooks, they would have died out a long time ago. "Humans have not engaged in hand-to-hand combat, at least not to the extent that would alter handedness frequencies worldwide, in modern times," says Holtzen. [35]

They have engaged in war, however. From the mano-a-mano of combat to the many-on-many of war, some of history's most heralded military leaders have been left-handed, and many of them persevered in their left-hand preference despite strong prejudice against the trait during the times in which they lived. The list, trotted out often in lefty lore, is an imposing one: Alexander the Great, Charlemagne, Ramses II, Tiberius, Commodus, Napoleon, three kings of England (Edward III, George II, George VI), Joan of Arc, Louis the XVI, General Patton, Fidel Castro, United States defense secretaries Robert McNamara and William Cohen, and the esteemed General Norman Schwarzkopf. [36]

A few common duties of leadership require skills that play to the strengths of a left-hander's brain. Military leaders must commandeer large numbers of forces across great distances, a task that requires an acute sense of geography, navigation, strategic planning, and "a strong grasp of the relationship of objects in different patterns of space," explains Harvard neuroscientist Albert Galaburda. [37] They also must anticipate problems and adapt if their plans and strategies "backfire." As a cadet in the United States Army, Norman Schwarzkopf tried to hold the trigger of an M-1 with his dominant left hand. The rifle was designed for right-handed shooters, and when he pulled the trigger, an explosion of hot gas ejected right into his face. Like most highly adaptive lefties, Schwarzkopf quickly learned to shoot from his right side. [38]

In addition to adapting in the moment, military strategy requires the ability to visualize a series of moves into the future, and to recognize geometric patterns among different arrangements. During the 1991 Gulf War, then-general Schwarzkopf was credited with the famous "left hook" military strategy, which avoided a traditional straight-up-the-middle assault on Saddam Hussein's troops in favor of a flanking maneuver that succeeding in defeating the Iraqi army in a matter of days.

It would be difficult, not to mention dangerous, for scientists to test whether military success can be linked to left-handedness, but scholars have found a connection between the trait and an aptitude most closely associated with military strategy: chess. Considered to be a formalized war game, chess players must move their army of knights, rooks, and pawns forward against an opponent's pieces in a series of strategic moves. "Chess requires visualizing what the board will look like several moves into the future," explains Harvard Medical School's Lee Cranberg. "Even more crucial is the ability to recognize patterns among the near-infinite arrangement of pieces. With a glance at the board, top players will recognize a pattern from games they've studied or played." Cranberg confirmed this in the late 1980s when he conducted a study of amateur and master chess players. His results found that nearly nineteen percent of players were left-handed, significantly higher than the left-handed population at large. [39]

The military arena is also compared frequently to the political arena, another domain where left-handedness is overrepresented, as seen in four recent United States presidents and several of their challengers. It's unclear whether the left-handed are more likely to be drawn to national politics, or whether they just persist and survive in disproportionately greater numbers, but according to Stanley Coren, the author of controversial *Left-Handed Syndrome,* they may be predisposed for the job: "Many left-handers actually have a profile that works very well for a politician. They tend to be more dominant rather than nurturing. And they tend to be a bit more pushy and a bit more cold." [40]

One person's "dominance" may be another person's coping skills. The negative stereotyping, while nothing new to lefties, may just dress them with the armor necessary for political life. Young sinistrals must battle resistance from well-meaning parents and teachers, and from the Bible to Webster's, they learn to live with insults: weak, gauche, cack-handed, cursed, ominous. "A child who overcomes the pressures of society . . . and persists in left-handedness, may develop a streak of stubbornness or an inclination to go against group pressure and accepted norms," Adam Blau, a psychiatrist and handedness researcher, told *Science News* in 1974. The resistance can breed a "willingness to go it alone in spite of other people's objections—as we have seen in Presidents Ford and Truman," Blau added. "Who would have thought that easy-going Truman would have dropped the bomb or that President Ford would have pardoned Mr. Nixon so soon?" [41]

As *Leading Minds* author Howard Gardner explains, "From early on, left-handers realize that they are different from other individuals, and that the world was not designed to serve them." A fiery determination to conquer a resistant terrain, and the ability to carefully navigate a world set up for others, could coax out a few other important leadership qualities. "One of the requirements for a leader—real or aspiring—is that he or she needs to be able to anticipate what is going to happen, to think about things in unconventional ways, to see himself or herself as different from the rest of the pack," Gardner concludes.

In more recent times, Blau might have pointed to the left-handed Bill Clinton and his determination to remain in office during his impeachment trial, despite pleas from members of his own party to step down, or the maverick senator from Arizona, John McCain, whose straight-talking refusal to consistently toe his own *rightward*-leaning party's line might be traced back to the developmental forces of a "stubborn lefty" childhood.

In addition to acquiring psychological defenses, the left-handed may develop a few practical skills that apply to modern-day campaigning. Gardner suggests that successful political candidates may be better at conceptualization and strategic planning, similar to the planning skills utilized by the great left-handed generals. "Often spatial or visual-spatial images help individuals when they are planning a campaign. It is probably not an accident that military leaders, who need to envision wide terrains, historically went into politics." [42]

Like Alexander the Great setting out to conquer the ancient world, today's national political contenders must envision the wide terrain that is the United States electoral map, with its fifty blue and red states, and conceptualize an optimal geographic strategy for a campaign. This is in addition to the ideological channel markers that must be navigated (zig to the right, zag to the center). Today's politicians must also be able to communicate in the visual mediums of television and the Internet, where they must rely on electronic "cave drawings" (photo ops) and the revelations of nonverbal expressions and gestures. He or she also must be able to communicate emotionally, using powerful metaphors and evocative symbols to drive home a message and reach people.

Bill Clinton is often singled out by historians, journalists, and fellow political leaders as the greatest politician to come along in a generation, largely as a result of his ability to relate to people on an emotional level. The left-hander understands how to use the nuances of speech, the full range of emotional intonations, and defusing doses of irony and humor, all of which could be attributed to the linguistic strengths gained from a greater integration of the left and right hemispheres. And like a rhesus monkey whose right cortex activates when it assesses the intentions of an approaching stranger, Clinton is also credited with an ability to read people, interpreting their sometimes hidden signals of anger, fear, sadness, or openness and responding accordingly, with his own wide-ranging facial expressions or his famously empathic "I feel your pain." Ultimately, a modern-day political candidate stands to be most successful if he or she is capable of integrating left and right, both within and without, as evidenced by the campaign talents of Ronald Reagan, George Bush Sr., John McCain, Bob Dole, Ross Perot, New York governor George Pataki, New York City mayor Michael Bloomberg, and of course, Bill Clinton, with the strategic counsel of a "war room" general such as George Stephanopoulos—all left-handers. [43]

Of course, it also helps if a political campaigner can keep an eye out for predators, whether in the form of political opponents or threats from other nations, so perhaps it helps to be left-eye dominant as well, like the chick scanning for hawks. Clearly, visual-spatial skills are not the only requirements for such important posts of leadership. It takes drive and perhaps even indignation to become a leader. "I think left-handed people tend to have a chip on their shoulder, and something they've got to prove," *Nightline's* Ted Koppel, a self-aware lefty, declared during the 1992 all-southpaw presidential showdown between George Bush Sr., Bill Clinton, and Ross Perot. Stanley Coren's self-reported personality study would seem to bear this out: "brazen" and "indignant," a few of the more common traits reported by left-handers in his 1995 survey, also match the traits required, or displayed, by many public officials. (Of course, one might want to take into account the possible response bias of a study in which individuals are brazenly willing to describe themselves using the traits listed on this survey.) [44]

It might also be the case that the physiological experience of tackling life from the other side can shape cognitive abilities, and in a way that might lead to fresh ideas and solutions, which are critical in public policy. In a 1995 study published in the *American Journal of Psychology*, the left-handed scored higher on tests of divergent thinking, a style of reasoning that leads to new concepts and ideas. "They are able to generate a multiplicity of solutions," concluded one of the study's authors. One could argue that Ross Perot's flat-tax platform, not to mention his decision to run for president twice in one election year, was the result of divergent thinking. It was Republican John McCain who championed campaign finance reform, and Democrat Bill Clinton who pushed through the outside-the-box solution to welfare reform—despite their respective parties' resistance to these ideas. Finally, one of the most admired left-handed leaders was not a warrior or a general or a president. Mahatma Gandhi, the spiritual and political leader who helped free India from British control, came up with a most divergent approach to change, that of nonviolent protest. His solution inspired civil-rights leaders like Nelson Mandela and Martin Luther King Jr., and today he is admired throughout the world as a symbol of peace and compassion.

Visionaries, Inventors, and Leonardo

When Leonardo da Vinci first conceived of a flying machine with flapping wings, in the 1480s, his designs were laughed at. Four centuries later, the Wright Brothers are credited for what was once dismissed as da Vinci's "fantastical" idea. Leonardo also conceptualized and drafted designs for history's first self-propelled vehicle. It would take another 431 years before fellow lefty Henry Ford produced the first Model T.

In 2005, an exhibit in Rome finally paid tribute to da Vinci's foresight, displaying models of his designs adjacent to their modern incarnations, many of which bear a close resemblance to his fifteenth-century visions. "Leonardo was beyond time," the curator of the exhibit, Carlo Barbieri, told Reuters. "He designed most of these things 500 years before they could even be built. It is impossible to say how he managed to imagine these things, he was too great a mind to comprehend." The curator does not specifically attribute da Vinci's greatness of mind to the fact that he was left-handed, but notes this was "part of the reason for his [mirror script] style of writing. But I think he largely chose to write this way to create a sense of mystery." His unusual writing style led some to consider him a wizard or mystic at the time, leading to many of the mysteries that shroud the man in books such as *The Da Vinci Code*. [45]

If, as is sometimes argued, challenging prevailing wisdom and indulging less-than-conventional thoughts (or those believed to be unconventional by the majority) are the seeds of invention, then the brain organization of a mixed left-hander may be the fertile ground for these seeds to take root. In 2002, Michael Corballis and a fellow New Zealander found a "systematic link" between ambidextrality, more equally developed hemispheres, and "magical ideation," a type of thinking that includes "irrational beliefs, delusions, and fantastical thoughts," but is often linked to creativity and invention. "I suspect Einstein is a case in point," says Corballis. Though Einstein's handedness is still a subject of some debate (most photos show him writing with his right hand), when the first anatomical study of his brain was conducted in 2000, neuroscientist Sandra Witelson found that the two hemispheres of his cortex, particularly the language centers, were more symmetrical than asymmetrical, a characteristic more prevalent in the left-handed. [46]

Connecting the dots between symmetrical brains and creative thinking is tricky, and connecting dots may be the best metaphor. One explanation is that a strongly specialized "lopsided" brain—which usually results from the presence of two right shift/dextral genes (R+ R+), but can also occur with a chance gene (R+ R-)—enhances the concentration of neurons in one hemisphere. Like a merger that centralizes power in one company's headquarters, the more concentrated hemisphere possesses a stronger "executive function," as Corballis explains. This hemisphere tends to be more ordered and tightly controlled; the chatter between the neurons is more likely to stay within the close confines of the central command center. In other words, it's easier to connect the dots—transmit signals between neurons. A more diffuse or symmetric brain organization leads to a less discrete executive function, a decentralized processing system that allows for more dispersed transmissions between neurons and more indirect associations, much like the ricochet of information that can lead to farther-fetched information in far-flung satellite offices.

Relying on these indirect associations and dispersed transmissions may have helped da Vinci become one of the broadest thinkers—and most prolific inventors—in history. In sketches and notes he scrawled in his left-handed mirror script, his thoughts zigzagged from philosophy to anatomy to astronomy to civil engineering to the mechanics of war. In one of his more densely written codices (today owned by Microsoft founder and fellow left-hander Bill Gates), da Vinci ruminates about geophysics "in a typically Leonardian modulation between the visionary and the practical," according to biographer Charles Nicholl. [47]

Da Vinci conceptualized and explained the principles for a parachute, a diving suit, a water-powered clock, a self-closing door, a folding bed, contact lenses, and a camera. He designed temples, hydraulic devices, bridges, watermills, armored cars, Archimedean screws, and solar-power panels. He sketched designs for a telescope that, when invented one hundred years later, enabled Galileo to confirm Copernicus's theory of a heliocentric solar system. He even designed a robotic lion for the king of France, Francis I, which could walk several steps. The king was so impressed he hired Leonardo as a counselor. [48]

Among the more significant examples of Leonardo's "divergent thinking" are his observations that upended the established scientific beliefs of his time. In the fifteenth century, people believed that the eye worked by emitting a stream of particles that reflected back off of objects, enabling us to see them. Leonardo dismissed this notion, and it led him to discover the optic nerve connecting the eye to the brain, and to postulate that light traveled at a certain velocity in straight lines. And while for centuries people had watched ripples spreading out from stones tossed into ponds, Leonardo was the first to recognize their significance. As he watched the ripples from two stones intersect without breaking up, he realized he was observing a shockwave that traveled through water, according to Michael White, author of *Leonardo: The First Scientist*. It was two hundred years before anyone else arrived at a wave theory of light, and this was Newton. [49]

But perhaps Leonardo's most fundamental talent was his ability to think in multiple dimensions and on multiple planes, relying on the spatial integration skills of his well-developed right hemisphere. According to education psychologist P. G. Aaron and historian Robert Crouse, da Vinci's brilliance was his ability to master and translate three-dimensional space in his sketches, paintings, and sculptures. He was the first to draw in a two-point perspective, sketching the Madonna and child by putting the room at an angle to the viewer's plane of sight, a "daring innovation not found in other Italian paintings at the time." As an engineer, he drafted blueprints for cranes, a horseless carriage, and bridges, and invented weapons that ranged from multiple mortars to siege cannons to armored tanks. All of these skills require an advanced mastery of visual-spatial integration. [50]

This ability to conceptualize objects in three dimensions and then transfer them to the two-dimensional plane is even more valuable in our present day, when information architects must conceive and develop complex multidimensional navigational systems, computer hardware and software, and Internet sites. "Mixed-handers may be at an advantage in being able to coordinate and integrate ongoing left and right hemisphere processing in complex, multidimensional tasks," suggests psychology professor Stephen Christman, referring to the large subset of left-handers who use their right hand for a few tasks such as throwing or cutting with scissors. For example, designing computer technology requires mediation between the right hemisphere's spatial planning and conceptual abilities and the left hemisphere's sequential reasoning and coding processes. Several great "mediated minds" are behind the two most successful computer companies: three of the four founders of Apple computers and the inventor behind Microsoft (again, Bill Gates) are all left- or mixed-handed. [51]

Sexual Selection, Bill Clinton, and Sarah Jessica Parker

Despite their many past and present contributions to society, the evolutionary fitness of the left-hander is still a subject of debate. Regardless of how many inventions and advancements lefties provide, in Darwinian terms they are still judged by their ability to provide the world with more lefties, and according to the theory of sexual selection, the infrequency of the trait must be a consequence of some disadvantage in the mating field.

As Chris McManus has argued, cultural bias against left-handedness was severe enough in the eighteenth and nineteenth centuries to limit the marriage prospects of those who possessed the trait. With fewer of them mating and passing along the genes, the left-handed dwindled to below five percent of the Western population by the early twentieth century. They redeemed themselves by quadrupling their numbers throughout the 1900s, at least in the United Kingdom, and McManus attributes this to a rise in the genes that result in left-handedness—thanks in part to the gradual lifting of the bias against the trait. Perceived as more eligible (or simply less "Neanderthal") mates, the left-handers were able to produce more offspring. [52]

Today, magazines bombard us with images of left-handed sex symbols, from Angelina Jolie to Sarah Jessica Parker, Brad Pitt to Prince Wills. And some of the most iconic sex symbols of all time, John Kennedy Jr., Robert Redford, Paul Newman, Steve McQueen, and Cary Grant, were all left-handed, and it's difficult to imagine any of them striking out in the local bar. Yet despite the apparent abundance of left-handed heartthrobs, there is statistical evidence that, historically, the southpaws did not have as many kids. [53]

What changed to make the gauche sinistrals more tempting? According to Michel Raymond and his colleagues, one answer lies in their athletic prowess. In 2003, three psychology professors from the University of Montpellier conducted a survey of college students and their sexual partners. The results were not surprising to anyone who has witnessed the aphrodisiacal powers of shaggy-haired Red Sox heartthrob Johnny Damon: competitive athletes of both sexes reported significantly more sexual partners than other students. The scholars added that the greater opportunities were not necessarily a function of attractiveness or body type. Instead, the athletes may have more opportunities to meet partners as they travel for competitions, or they may have more sexual motivation to begin with, which led them to pursue sports more rigorously. "Why left-handedness persists has been a puzzle because of evidence of associated fitness costs," they conclude, re-cycling the old "syndrome" theories from the 1980s and 1990s. "It would seem that some countervailing benefit, probably frequency-dependent, is required to maintain [left-handedness], and the results of our study suggest that in light of the disproportionate prevalence of left-handers among athletes, that benefit may reside, at least in part, in the sexual success of athletes." [54]

In cultures where sexual finesse ranks high on the list of "mate" criteria, the lefties may even have a few tactical advantages, particularly when it comes to the "interpersonal skills" required during foreplay. In some African societies, males are expected to lie on their right sides during intercourse, and use their left hand to stimulate their partners before sex, and in similar situations across all societies, if a man—or woman—finds him- or herself lying opposite a sexual partner who is right-handed, southpaw skills can come in handy. [55]

Michelangelo, Mozart, and McCartney

Athletic abilities and sexual techniques are not the only prowesses that attract potential mates. Artistic talents have also been selected by the members of both sexes throughout evolution, and the scholarly research is rich with findings that the left-handed have a few special capacities in these departments as well. In fact, throughout the animal kingdom, "artistic talents" prove to be a powerful way to woo the opposite sex. In his book, *The Mating Mind*, evolutionary psychologist Geoffrey Miller observes the show-off mating techniques of male bowerbirds, which include constructing elaborate nests to attract females, decorating them with brightly colored flowers, feathers, shells, and pebbles. The females tour the various displays, choose the one they find most attractive, and copulate with its creator. They then leave to raise their brood in a plain nest of their own creation. [56]

Of course, humans have gone to similar lengths to display their aesthetic abilities, from cave paintings to skin decorations to elaborate huts and castles. According to Miller, the money and effort spent on clothes, makeup, or elaborate architectural "nests" can be viewed as modern-day extensions of this mating strategy. Rooted in this evolutionary drive is the assumption that the more pleasing the décor, the more "fit" the creators. As Marian Annett already theorized, the left-handed are more likely to have a net sum total of "manual skills," but they are also known to have some built-in talents when it comes to constructing intricate nests. In the majority of lefties, the brain is set up such that visual-spatial abilities and three-dimensional planning are adjacent to the headquarters for the drawing hand, and as we've learned from da Vinci, this arrangement can facilitate the translation of visions and concepts onto paper. Studies have found that left-handers make up a remarkably high proportion of architecture-school faculty members (twenty-nine percent) and students (twenty-one percent). The tendency toward greater cross-talk between the brain's hemispheres could also be at work in these numbers. As Steve Christman explains, "I suspect that mixed-handed architects may be better able to find suitable compromises between the overall form of a building (a right hemisphere function) and specific details required by the building codes, such as ventilation ducts (which would be more of a left hemisphere function)." [57]

Musical virtuosity, another show-off talent used to woo potential mates, could be added to the left-handed repertoire, offsetting the weight of stigma and right-shifting evolutionary forces. The human equivalent of birdsong—songwriting—may come more easily to certain subsets of lefties, according to Christman. "I strongly suspect that mixed-handers may have an advantage in putting the right words to the right melody. Words are processed in the left hemisphere (in the majority of this group) and melodies, harmonies, and affective tone in the right. Strong handers may be good at just the words or just the melody, but I suspect that they may not be as good at putting them together." Christman points to evidence in the surprising number of talented songwriters from recent decades who were left-handed, among them Paul McCartney, Paul Simon, Carly Simon, Jimi Hendrix, Bob Dylan, and Cole Porter. He even adds controversial white rapper Eminem to the list, "because, as someone who cut his teeth on Dylan songs, I think Eminem is perhaps the most clever wordsmith to come along since Dylan." A few other modern musician-songwriters can be added to the list, including Sting, David Bowie, Phil Collins, The Talking Heads' David Byrne, Greg Allman, Led Zeppelin's Robert Plant, REM's Michael Stipe, Nirvana's Kurt Cobain, Lauryn Hill, Seal, Celine Dion, and Annie Lennox. Christman, himself a left-handed songwriter, explains that "words and melodies routinely come to me spontaneously at the same time. I've almost never written a song where I start with just lyrics and try to write a melody (or vice versa)." [58]

Mastery of musical instruments benefits from two skilled hands, and as noted earlier, the left-handed tend to have stronger nondominant hands, as well as an innate ability to integrate the roles of the two hemispheres. Both attributes are helpful when it comes to playing instruments that require coordination between the hands, perhaps explaining the disproportionate number of left-handers among the ranks of professional wind and string musicians. [59]

While athletic and artistic prowess may have helped the left-handed overcome some of the cultural biases stacked against them, for others their very freakishness was a draw in itself. As evolutionary theorist Miller explains, some early humans developed sexual preferences for unpredictability and "interesting, funny mates." He calls this "the dazzle factor," a sexual survival skill that comes into play in species that no longer need to blend in to stay alive. Our brains became endowed with mechanisms that allowed for "randomizing social behaviors," says Miller. So if the left-handed members of the species lacked the standard right-handed abilities or predictable cerebral arrangements, their less conventional attributes might have been a draw after all—at least before they became stigmatized by religious symbolism and the emphasis on social conformity. [60]

"The ability to produce amusingly unpredictable romantic behavior" helped lure new mates, says Miller, who suggests that "neophilia," or the attraction to novelty, could have been driven by mate choice as well. He points to research by Cornell University's Meredith Small, who has studied female sexual behavior in primates and concluded, "The only constant interest seen among the general primate population is an interest in novelty and variety." According to Small, "Although the possibility of choosing for good genes, good fathers, or good friends remains an option open to female primates, they seem to prefer the unexpected." [61] Miller offers examples from research on birds. "Large song repertoires, as seen in some bird species like sedge warblers and nightingales, allow birds to produce the appearance of continuous musical novelty." Darwin himself observed in *Descent of Man* that for birds, "mere novelty, or slight changes for the sake of change, have sometimes acted on female birds as a charm, like changes of fashion with us." [62]

In humans, evidence of neophilia can be found in our novelty-driven economy, according to Miller, who points to the industries of news, fashion, film, television, publishing, psychoactive drugs, and music. "It seems likely that our hominid ancestors were highly appreciative of novelty, and that this neophilia spilled over into mate choice, where it favored not so much a diversity of sexual partners, but selection of highly creative partners capable of generating continuous behavioral novelty throughout the long years necessary to collaborate on raising children." The challenge, according to Miller, was to convince mating prospects that they would be entertained over the course of a long relationship, "so they don't get bored and incur the maladaptive costs of separation and searching again." Partners who offered more creativity and cognitive variety may have had longer, more productively successful relationships, adds Miller. So, while natural selection might have favored those who could acquire accurate, survival-enhancing knowledge, sexual selection would have favored minds "prone to inventing attractive, imaginative fantasies." [63] (Consider *Alice in Wonderland*, for example, which was created by left-handed author Lewis Carroll, and the dreamscape river adventures of Huck Finn and Tom Sawyer, spun from the left-handed imagination of Mark Twain.)

* * *

Today, the left-handed can be found in every country, race, religion, and occupation. They are numbers people and word people, and numbers people who like words. They are entrepreneurs and moguls, and entrepreneurs who become moguls (Bill Gates). They are rational-minded leaders and free-spirited artists, and free-spirited leaders who can also be rational-minded artists (think Oprah). They are comedians and thespians, lawyers and psychiatrists, athletes and architects, specialists and generalists. They are "left-brained" lefties and "right-brained" lefties, and those who fall somewhere in between. They are different from their right-handed brethren, and often different from each other. In fact, if there is one generalization that can be made about them, it is that they are more diverse and unpredictable than right-handers—in the organization of their brains and how they experience the world.

"It's an advantage for human society to have cognitive and emotional diversity that make people have different likes and dislikes, be good at different things and weak at others," adds Jerre Levy, professor of psychology at the University of Chicago. "You can imagine a whole range of people, some of whom are extreme cognitive specialists, let's say geniuses in mathematics but limited in everything else. Fortunately we now have a society in which that person can make a contribution to intellectual knowledge. We should leave him alone and let him do his math. And other people can grow his food and build his house and educate his children and so forth.

"And there may also be cognitive generalists who are not extremely specialized at any one thing. So they may not be as good a mathematician as this guy or as great a composer as that woman or as wonderful a painter as this artist. But they're nonetheless cognitive generalists that could, to a certain extent, integrate all of these different things. If they could serve as an integrative force in this society among people who were specialists, they would bring a little understanding of everything," says Levy. [64]

Some of history's most celebrated "integrators" were either left- or ambi-handed jacks-of-all-trades. Alexander the Great was a warrior, visionary, and world leader who, in thirty-three years, created the first modern empire, designed the first planned city, invented snorkel diving, and cut the Gordian knot. Ben Franklin was a printer, scientist, author, inventor, diplomat, and Founding Father. Nineteenth-century prodigy Charles Sander Pierce, who founded semiotics, was an accomplished mathematician, astronomer, and logician who invented the philosophical term *pragmatism*. Albert Schweitzer, the Frenchman who won the Nobel Peace Prize for his missionary work in Africa, was also a musician, a physician, and a theologian.

"In every generation throughout history, we have had a subset of people who really are outside the box," adds psychologist David Holtzen. "We have to have that." [65]

"The left stuff" is all of that out-of-the-box stuff. The trait grants those who possess it a tendency toward more flexible thinking, the psychological freedom to question orthodoxies, and the cognitive facility to update beliefs. "In the end, my view is that society as a whole benefits from having members of both handedness groups," says Christman, "by having a balance between the 'change-resistant' strong handers and the 'I'll give anything the benefit of the doubt' mixed handers. The mixed-handers may serve as society's guinea pigs, trying out controversial new ideas, while the strong handers serve as society's ballast, ensuring that we do not fall headlong into new, untested ideas that may not stand the test of time." [66]

So while their minority status and "backward" orientation may create obstacles for the left-handed, these challenges have also helped them, and the species, survive.

Afterword

Human salvation lies in the hands of the creatively maladjusted
—MARTIN LUTHER KING, JR.

It has been nearly twenty years since a flurry of flawed reports first proposed that left-handedness was a pathology, tagging those who possessed the trait with "decreased fitness" and scaring them with predictions of a shortened life span. More than a decade after these claims were debunked, the myth of the dying and disordered lefty still persists, adding to the bias that has plagued the port-sided throughout history.

In spite of all the bad press, perhaps the left-handed should be grateful for all the threats and labels they have had to endure. While sensational claims make newspaper headlines, the weight of science is on their side, as current research tells a tale not of shortened lives and mysterious ailments, but of strengths and surging numbers. As the evidence to support their evolutionary role mounts, so does evidence that the hurdles they have faced only succeeded in making them stronger. The corkscrews and camcorders and computer mice that have thwarted them have also spurred on their brains. And like this struggle to adapt to a right-handed world, facing down the adversaries and adversities—the insults and indignations and knuckle-rapping—may also be contributing to the left-handers' triumphs. "There are a large number of individuals who, when faced with a number of stressors, are driven to become more successful," says Coren. "Some even achieve greatness."

Up against near universal resistance, the left-handed have become a resourceful and resilient bunch. They're a group that perseveres in every culture around the globe regardless of cultural oppression. They throw curveball strikes, lead armies, and guide countries. Run multibillion-dollar companies and multichild households. They may be the misfits of standardization, but they have prevailed in every arena. And the more scientists make forays into this subject, the more the stigmas will disappear.

In the near future, it's likely that the identity and location of the gene or genes responsible for handedness and brain dominance will be found. As is the case with all genetically determined human characteristics, this will likely open the door to new issues surrounding left-handedness. If people are able to choose the handedness of their children, will the myths of premature death or lingering social stigma lead them to choose to bear only right-handed children? By making the decision to produce a right-handed child, would these parents also be making the decision to deprive the world of another da Vinci? Another Babe Ruth? A future president?

While we can only hypothesize about what the discovery of the true left-handedness gene would mean socially, a better question than what we will lose from this pursuit of laterality's answers is what we will gain. Today, scientists can agree that left-handedness is not an unintended consequence. It is one of Darwin's variations, a trait as intrinsic to humanity as all of our evolved traits, and important enough to have survived since early man, perhaps even before—precisely because it rests at the heart of Darwin's theory: that the variations within a species, diversity itself, are what keep us thriving. After all, the French, who gave us "gauche" for left, also gave us another expression: "vive la difference."

Stereotypes and stigmas are stubborn, and convincing a hostile world that lefties are not debilitated righties will take time. The essential matter is that scientists continue to explore and discover along this frontier; that they delve deeper into the recesses of the human brain, searching for the answers that our hands can provide about ourselves and our humanity.

This will be neither the last word nor the last book written on lefties. The state-of-the-science described in these pages is offered not as conclusive evidence of any particular theory, but as celebration of the new breakthroughs that emerge almost daily, allowing us to understand not just how our hands operate, but how their uneven skills precisely relate to our brains' functions. Our hands are the waving, shaking, and gesturing windows into our cerebral arrangements. Their movements reflect not only our intricate cross-wiring, but also our complex ability to communicate. So the search for knowledge about laterality is important not only because it dissolves the long-standing misconceptions about lefties, but also because it forms a crucial piece of the human evolution puzzle, bringing us to a fuller, more complete version of ourselves. Even if it does come from out of left field.

Appendix

Handedness Surveys

The Edinburgh Handedness Survey

Which hand do you prefer, and do you ever use your other hand, for the following tasks:

1. Writing
2. Drawing
3. Throwing
4. Cutting with scissors
5. Cutting with a knife (no fork)
6. Spooning something out
7. Teethbrushing
8. Sweeping (upper hand on broom)
9. Striking a match
10. Opening a box lid

Which eye do you use to look through a camera lens?

Which foot do you use to kick a ball?

Source: Oldfield, R. C. (1971) The assessment and analysis of handedness: the Edinburgh inventory. *Neuropsychologia*, 9(1): 97–113.

Take the test and get an instant score:
http://porkpie.loni.ucla.edu/LabNotes/edinburgh.html

The Annett Hand Preference Questionnaire

Please indicate which hand you habitually use for each of the following activities by writing R (for right), L (for left), E (for either).*

1. To write a letter legibly?

2. To throw a ball to hit a target?

3. To hold a racket in tennis, squash, or badminton?

4. To hold a match whilst striking it?

5. To cut with scissors?

6. To guide a thread through the eye of a needle (or guide needle onto thread)?

7. At the top of a broom while sweeping?

8. At the top of a shovel when moving sand?

9. To deal playing cards?

10. To hammer a nail into wood?

11. To hold a toothbrush while cleaning your teeth?

12. To unscrew the lid of a jar?

If you use the *right hand for all of these actions*, are there any one-handed actions for which you use the *left hand*? Please record them here:

_____

If you use the *left hand for all of these actions*, are there any one-handed actions for which you use the *right hand*? Please record them here:

_____

*This instruction was omitted in some versions with the intention of discouraging "E" responses.

Source: Annett, M. (1970). A classification of hand preference by association analysis. *British Journal of Psychology, 61,* 303–321.

Endnotes

Chapter 1

1 Gilbert, A. N., and Wysocki, C. J. (1992). "Hand preference and age in the United States." *Neuropsychologia, 30,* 601–608. Searleman, A., and Porac, C. (2003). "Lateral preference profiles and right shift attempt histories of consistent and inconsistent left-handers." *Brain and Cognition, 52,* 175–180. McManus, Chris (2002). *Right Hand, Left Hand. The Origins of Asymmetries in Brains, Bodies, Atoms and Cultures.* London: Weidenfeld and Nicolson, p. 208. Leake, J., and Grimston, J. "Left-handed prodigies stage a renaissance," *Sunday Times* (London), March 17, 2002.

2 Hertz, R. (1960) *Death and the Right Hand.* Aberdeen: Cohen and West, pp. 93–103. Domhoff, William (1969). "Why do they sit on the king's right in the first place?" *Psychoanalytic Review,* Winter 1969–1970, 588–589. Fincher, Jack (1977). *Sinister People: The looking-glass world of the left-hander.* New York:
G P Putnam's Sons, pp. 30–39.

3 Ward, Jeannette, Memphis State University. Interviews with author (2001, 2004).

4 Restak, Richard M. "Do we want a left-handed president?" *The Washington Post,* December 11, 1988. Coren, Stanley (1992). *The Left-Hander Syndrome.* New York: The Free Press, p. 210. Coren, S., and Halpern, D. F. (1991). "Left-handedness: A marker for decreased survival fitness." *Psychological Bulletin, 109*(1): 90–106. Halpern, D. F., and Coren, S. (1988). "Do right-handers live longer?" *Nature, 3*(33):

213. Hilts, Philip. "A sinister bias: New studies cite perils for lefties." *The New York Times*, August 29, 1989.

5 Coren (1992). Okie, Susan. "Bush's thyroid condition diagnosed as Graves' disease." *The Washington Post*, May 10, 1991.

6 Connor, Steve. "Leading with your left may shorten your life." *The Independent* (London). March 3, 1991. Gladwell, Malcolm. "Left-handers die younger, study says." *The Washington Post*, April 4, 1991. Roark, Ann C. "Death rates studied: Is the world really safe for lefties?" *The Los Angeles Times*, September 14, 1989.

7 "Stats from left field." *The Washington Post*, April 20, 1991.

8 Porac, Clare, et al. (1998). "Illness and accidental injury in young and older adult left- and right-handers: Implications for genetic theories of hand preference." *Developmental Neuropsychology, 14*(1): 157–172. Porac, C., and Searleman, A. (2002). "The effects of hand preference side and hand preference switch history on measures of psychological and physical well-being and cognitive performance in a sample of older adult right and left handers." *Neuropsychologia, 40*, 2074–2083.

9 Bumiller, Elisabeth. "Council urged to end a most sinister bias." *The New York Times*, June 22, 2000.

10 Elston, Laura. "Shorter, tougher life for left-handed." *The Independent* (London), August 13, 2000. Berwick, Isabel, and Leigh, Adam. "Debate: Campaign in the United States is pushing for anti-discrimination laws to protect left-handed people." *The Independent* (London), April 11, 1999.

11 Berwick and Leigh, ibid.

12 Marks, J. S. and Williamson, D. F. (1991). "Left-handedness and life expectancy." *New England Journal of Medicine, 325;* 1042. Rothman, K. J. (1991). "Left-handedness and life expectancy." *New England Journal of Medicine, 325;* 1041. Murphy, James J. "Long live lefties." *The New York Times*, February 15, 1992. Salive, M. E., Guralnik, J. M., and Glynn, R. J. (1993). "Left-handedness and mortality." *American Journal of Public Health*, 83, 265–267. Harris, Lauren J. (1993). "Do left-handers die sooner than right-handers? Commentary on Coren and Halperin's (1991) 'Left-handedness: A marker for decreased survival fitness.'" *Psychological Bulletin, 114*(2): 203–234. Annett, M. (1993). "The fallacy of the argument for reduced longevity in left-handers." *Perceptual and Motor Skills, 76*, 295–298. Hicks, R. A. et al. (1994).

"Do right-handers live longer? An updated assessment of baseball player data." *Perceptual and Motor Skills*, 78, 1243–1247. Murray, David. "Exit, stage left." *The National Post* (Toronto, Canada), August 18, 2000.

13 Rosemary West's site, www.rkwest.com. Author outreach to research scientists.

14 Peters, M., and Perry, R. (1991). "No link between left-handedness and maternal age and no elevated accident rate in left-handers." *Neuropsychologia*, 29, 1257–1259. Murphy, J. ibid. Shute, Nancy. "Southpaw Reprieve." *U. S. News & World Report*, October 9, 2000. Porac (1998). et al.

15 Holder, M. K. (2002). "Gauche! Left-handers in society." www.indiana. edu/~primate/lspeak.htm. Fincher, ibid. Corbis and AP archival photo research.

Chapter 2

1 Englund, Will. "In Russia, left isn't quite right; Handedness: The official Moscow line is that lefties are OK, but suspicion of those who are different persists from the old Soviet days." *The Baltimore Sun*, March 27, 1998.

2 McManus, Chris. (2002). *Right Hand, Left Hand. The Origins of Asymmetries in Brains, Bodies, Atoms and Cultures*. London: Weidenfeld and Nicolson, p. 208.

3 Hertz, R. (1960). *Death and the Right Hand*. Aberdeen: Cohen and West, pp. 102–103.

4 Webster's Dictionary, 1996. Roget's College Thesaurus, 1995. Fincher, Jack. (1977). *Sinister People: The Looking-Glass World of the Left-Hander*. New York: G P Putnam's Sons, p. 37. Wile, I. S. (1934). *Handedness: Right and Left*. Boston: Lethrop, Lee and Shepard, pp. 40, 334. de Kay, James T. (1966) *The Left-Handed Book*. M. Evans and Company, Inc.

5 Hertz, R. p. 103.

6 Fabbro, F. (1994). "Left and right in the Bible from a neuropsychological perspective." *Brain and Cognition*, 24, 161–183.

7 Goddard, Dwight. *The Teaching of Buddha, the Buddhist Bible*. Santa Barbara, CA: J. F. Rowny Press. 1934. Mcmanus, p. 32. Anonymous

"Left-hand path and right-hand path." Wikipedia http://en.wikipedia.org.

8 Fabbro, ibid.

9 Hecaen, H., and Sauguet, J. (1971). "Cerebral dominance in left-handed subjects." *Cortex, 3* (71): 125.

10 Fincher, pp. 33–34.

11 Fincher, ibid. Hertz, pp. 97–99.

12 Hertz, pp. 102–103.

13 Hertz, p. 107.

14 Davies, R. Trevor. (1947). *Four Centuries of Witch-Beliefs, with Special Reference to the Great Rebellion.* London: Methuen and Co. Ltd. Boyer, P., and Nissenbaum, S. (Eds). "The Salem witchcraft papers." Volume 1: Verbatim transcripts of the legal documents of the Salem witchcraft outbreak of 1692. Electronic Text Center, University of Virginia Library.

15 Pegrum, Juliet. (2003). *Hatha Yoga: The Complete Mind and Body Workout.* New Delhi: Sterling Publishing Co. Scholem, Gershom. (1969). *On the Kabbalah and Its Symbolism.* New York: Schocken Books.

16 Hertz, p. 102.

17 Brown, Dan. (2003). *The DaVinci Code.* New York: Doubleday, p. 125.

18 Hecaen and Sauguet. Hertz, p. 98.

19 Hertz, p. 102.

20 De Kay, p. 44. Lane, E. W. *An Account of the Manners and Customs of the Modern Egyptians.* London. 1836. J. K. Dent & Sons Ltd: New York EP Dutten & Co.

21 Shute, Nancy. (December 1994). "Life for lefties: From annoying to downright risky." *The Smithsonian, 25* (9): pp. 130–143.

22 Lombroso, Cesare. (1876). *The Delinquent Male. From Coren:* Lombroso C. *North American Review*, 1903. (177): 440–444. Rutledge, L., Donley, R., and Bennett, J. (1992). *The Left-Hander's Guide to Life.* New York: Penguin.

23 Richardson, Bob. "The invaluable lefty; Game's funniest characters left 'em laughing." *The Boston Globe*, March 31, 1989.

24 Barnett, K. J. and Corballis, M. C. (2002). "Ambidexterity and magical ideation." *Laterality, 7*, 75–84. Author correspondence with Corballis, 2001.

25 Harris, Lauren J., correspondence with author (2004). Peters, Michael, correspondence with author (2004). Delisi L. E. et al. (2002). "Hand

preference and hand skill in families with schizophrenia." *Laterality,* 7(4) 321–332.

26 Harris, ibid.

27 Coren (1992). pp. 66–67. Porac, Clare, correspondence with author. (2004). Shipp, E.R. "The pols and zealots misuse the Bible." *The New York Daily News,* February 29, 2004.

28 Culberston, Debbie. (Fall 2001). "A Conversation with Derek Evans." *Exchange.* The United Church of Canada, October 3, 2002.

29 Grimston, J., and Leake, J. "Left-handed prodigies stage a renaissance." *Sunday Times* (London), March 17, 2002. Rutledge, L. W., and Donley, R. (1992). *The Left-Hander's Guide to Life.* New York: Penguin. Transcript: "The Listener: Strike a Chord; This Week's Selection of the Best of BBC Radio." London *Independent,* October 24, 1999.

30 Schugurensky, Daniel. "History of Education: Selected Moments of the Twentieth Century." The Ontario Insititute for Studies in Education of the University of Toronto (OISE/UT). Author correspondence with Porac (2004). BaseballAlmanac.com. The Baseball Hall of Fame.

31 Spock, Benjamin. (1946). *The Common Sense Book of Baby and Child Care.* New York: Duell, Sloan and Pearce. Author interview with Lauren J. Harris (2004).

32 Healey, Jane M. *Loving Lefties: How to Raise Your Left-handed Child in a Right-Handed World.* New York: Pocket Books. 2001. pp. 48-49. Porac, C. (2003). "Hand skill and preference in switched left-hand writers." *Canadian Psychology,* 44 (2a): 152.

33 Perelle, I. B., and Ehrman, L. (1994). "An international study of human handedness: The data. *Behavior Genetics,* 24, 217–227. Payne, M. A. (1987). "Impact of cultural pressures on self-reports of actual and approved hand use." *Neuropsychologia,* 25, 247–258. Harris, L. J. (1990). "Cultural influences on handedness: Historical and contemporary theory and evidence." In S. Coren (ed.), (1990). *Left-handedness: Behavioral implications and anomalies.* Amsterdam: North-Holland, Elsevier, pp. 195–258. Shimizu, A., and Endo, M. (1983). "Handedness and familial sinistrality in a Japanese student population." *Cortex,* 19, 265–272. Author correspondence with Peters, Harris (2004).

34 Porac, C., and Searleman, A. (2002). "The effects of hand preference side and hand preference switch history on measures of psychological

and physical well-being and cognitive performance in a sample of older adult right and left handers." *Neuropsychologia, 40,* 2074–2083. McManus, pp. 269–270. Craig's List: http://forums.newyork.craigslist. org/? forumID=39. Porac, C. et al. (ed.) 1990.

35 McManus, Ibid. Healy, Jane M. (2001). *Loving Lefties: How to Raise Your Left-Handed Child in a Right-Handed World.* New York: Pocket, pp. 67–68.

36 ChangeThatsRightNow.com, "Left-Handed Phobia Clinic." Domhoff, William. (1969). "Why do they sit on the king's right in the first place?" *Psychoanalytic Review,* Winter 1969–1970, 588–589.

Chapter 3

1 Microsoft Communications Department, author correspondence (2001).

2 Bishop, D.V.M. et al. (1996). "The measurement of hand preference: A validation study comparing three groups of right-handers." *British Journal of Psychology,* 87(2): 269. Interviews with Jeannette Ward (2001, 2004).

3 Smith, Barry D., Meyers, M. B., Kline, R. (1989). "For better or for worse: Left-handedness, pathology and talent." *Journal of Clinical and Experimental Neuropsychology,* 11(6) 944–958.

4 Porac, C. (1993). "Are age trends in adult hand preference best explained by developmental shifts or generational differences?" *Canadian Journal of Experimental Psychology.* 47;4: 697–713; Steenhuis, R. E. et al. (1990). "Reliability of hand preference items and factors." *Journal of Clinical Experimental Neuropsychology,* 12(6) 921–930.

5 Creamer, Robert W. 1974. *Babe: The Legend Comes to Life.* New York: Simon and Schuster.

6 Reagan anecdotes: Rosenbaum, David E. "On left-handedness, its causes and costs." *The New York Times,* May 16, 2000. Others: "Baseball as America" exhibit at Smithsonian, 2004. Haikola, Mauri. (www.stekt. oulu.fi/~mjh/lefties.html). Anything Left-Handed.co.uk.

7 Annett, Marian, correspondence with author (2001).

8 Witelson, S. F. (1985). "The brain connection: the corpus callosum is larger in left-handers." *Science,* 229(4714) 665–668.

9 Annett, M. (1992). "Spatial ability in subgroups of left- and right-

handers." *British Journal of Psychology, 83,* 493–515. (See also: Annett, M. (2002). *Handedness and Brain Asymmetry: The Right Shift Theory.* Hove, UK: Psychology Press.

10 Annett, M. (1970). "A classification of hand preference by association analysis." *British Journal of Psychology, 61,* 303–321.

11 Oldfield, R. C. (1971). "The assessment and analysis of handedness: The Edinburgh inventory." *Neuropsychologia, 9*(1) 97–113.

12 Bishop (1996). Steenhuis et al (1990).

13 Peters, interview with author (2001).

14 Bishop et al. (1996).

15 Klar, A. J. S. (2003). "Human handedness and scalp hair whorl direction develop from a common genetic mechanism." *Genetics.* Vol. 165, 269-276, September 2003.

16 Needlman, Robert, M.D., F.A.A.P., American Online Parenting, August 2001. Witelson, S. F. and Wazir, P. "Left hemisphere specialization for language in the newborn." *Brain,* 1973(96) 641–646.

17 Michel, G. F. (1983). In Young, G. et al (Eds). *Manual Specialization and the Developing Brain.* New York: Academic Press, pp. 33–70. McManus, I. C. et al. (1988). "The development of handedness in children." *British Journal of Developmental Psychology, 6,* 257–273. Harris, L. J. (1983). "Laterality of function in the infant: Historical and contemporary trends in theory and research." In G. Young et al. (eds). *Manual Specialization and the Developing Brain.* New York: Academic Press, pp. 177–247.

18 Harris (1983). Also Harris interview with author (2004). Needlman (2001).

19 Tosches, Rich. "Lefties: Make no mistake, it's a right-hander's world. Except in sports—where being out of step can put a player in sync." *Los Angeles Times,* December 4, 1986. Baseball Hall of Fame.

20 Laland, Kevin, correspondence with author (2002).

21 Christman, S. D., and Propper, R. E. (2001). "Superior episodic memory is associated with interhemispheric processing." *Neuropsychology, 15,* 607–616. Author correspondence with Christman (2001, 2004).

22 Peters, M. (1995). "Handedness and its relation to other indices of cerebral lateralization." In R. Davidson and K. Hugdahl (Eds.), *Brain*

Asymmetry. Cambridge, MA: MIT Press. Author correspondence with Peters (2001, 2004).

23 Smith, Barry et al. (1989). Hicks et al. (1993). "Handedness and accidents with injury." *Perceptual and Motor Skills, 77*(3): 1119–1122. Correspondence with Hicks (2000).

24 Peters, correspondence with author (2004).

25 Perelle, I. B., and Ehrman, L. (1994). "An international study of human handedness: The data." *Behavior Genetics, 24,* 217–227. Payne, M. A. (1987). "Impact of cultural pressures on self-reports of actual and approved hand use." *Neuropsychologia, 25,* 247–258. Harris, L. J. (1990). "Cultural influences on handedness: Historical and contemporary theory and evidence." In S. Coren (Ed.). *Left-handedness: Behavioral implications and anomalies* (pp.195–258). Komai, T., and Fukuoka, G. (1934). "A study of the frequency of left-handedness and left-footedness among Japanese school children." *Human Biology, 6,* 33–42. Author correspondence with Porac (2004).

26 Peters, M. (1995). Peters interviews with author (2004).

27 Komai and Fukuoka, ibid. Kang Y., and Harris L. J. (2000). "Handedness and footedness in Korean college students." *Brain and Cognition, 43*(1–3): 268–274.

28 Porac, C, et al. (1998). "Illness and accidental injury in young and older adult left- and right handers: Implications for genetic theories of hand preference." *Developmental Neuropsychology, 14*(1): 157–172. Porac, C. (1996). "Attempts to switch the writing hand: Relationships to age and side of hand preference." *Laterality, 1*(1), 35-44. Porac, Ibid.

29 Searleman, A and Porac, C. (2003). "Lateral preference profiles and right shift attempt histories of consistent and inconsistent left-handers." *Brain and Cognition, 52* (2003) 175–180.

30 Christman, S. D., and Propper, R. E. (2001). "Superior episodic memory is associated with interhemispheric processing." *Neuropsychology, 15,* 607–616. Author correspondence with Christman (2001, 2004).

31 Peters, M. (1988). "Footedness: asymmetries in foot preference and skill and neuropsychological assessment of foot movement." *Psychological Bulletin, 103*(2): 179–192. Reiss, M., et al. (1999). "Laterality of hand, foot, eye and ear in twins." *Laterality, 4*(3): 287–297.

32 McManus, I. C. et al. (1999). "Eye-dominance, writing hand and throwing hand." *Laterality,* 4(3): 173–192. Coren, Stanley. (1992). *The Left-Hander Syndrome.* New York: The Free Press (pp. 29–31). Feisig, Glenn, American Sports Medicine Institute, interview with author (2004).

33 Coren, p. 31.

34 Ibid.

35 Ibid.

Chapter 4

1 Springer, Sally P., and Deutsch, George. (1985). *Left Brain, Right Brain: Perspectives in Cognitive Neuroscience.* New York: WH Freeman and Company, p. 39.

2 Bryden, M. P. (1982). *Laterality: Functional Asymmetry in the Intact Brain.* New York: Academic Press, p. 170.

3 Edwards, Betty. (1979). *Drawing from the Right Side of the Brain.* Los Angeles: J. P. Tarcher, pp. 23–25.

4 Galaburda, Albert M. (1996). "Anatomic Basis for Cerebral Dominance." In Davidson, R. J., and Hugdahl, K. (1998). *Brain Asymmetry.* Cambridge, MA: MIT Press, pp. 51–73. Bryden, ibid.

5 Society for Neuroscience, "Brain Facts." http://apu.sfn.org/content/Publications/BrainFacts/index.html. University of Washington, Department of Neuroscience, "Brain Basics." http://faculty.washington.edu/chudler/introb.html.

6 Harrington, Anne. (1996). "Unfinished business: Models of laterality in the nineteenth century." In Davidson, R.J. and Hugdahl, K. *Brain Asymmetry.* Cambridge, MA: MIT Press, p. 13.

7 Springer and Deutsch, pp. 33–58.

8 Ibid.

9 Bogen, J. E. (1977). "Hemisphere specialization." In Wittock, M. C. (Ed). *The Human Brain.* Prentice Hall. Davidson and Hugdahl (1996), ibid. Ornstein, Robert. (1997). The Right Mind. Making Sense of the Hemispheres. New York: Harcourt Bruce. pp. 65–70.

10 Springer and Deutsch, p. 39.

11 Ibid.

12 Springer and Deutsch, pp. 48–51.

13 Springer and Deutsch, ibid. Ornstein. p. 66.

14 Springer and Deutsch, ibid.

15 Kimura, D. (1993). *Neuromotor mechanisms in human communication*. New York: Oxford University Press. Bellugi et al. (1983). "Brain organization for language: Clues from sign aphasia." *Human Neurobiology, 2,* 155–170.

16 Rotenberg, V. S., and Arshavsky, V. V. (1991). "Psychophysiology of hemispheric asymmetry: The 'entropy' of right hemisphere activity." *Integrative Physiological and Behavioral Science, 26,* 183–188.

17 Ibid.

18 Ibid.

19 Ramachandran, V. S. and Blakeslee, Sandra. (1999). *Phantoms in the Brain: Probing the Mysteries of the Human Mind*. New York: HarperCollins, pp. 129–131.

20 Ibid, p. 130.

21 Ibid, pp. 135–136.

22 Bourgeois, M., Christman, S., and Horowitz, I. (1998). "The role of hemispheric activation in stereotyping versus individuation: Evidence for two separate subsystems in person perception." *Brain and Cognition, 38,* 202–219.

23 Davidson, Richard. (1996). "Cerebral asymmetry, emotion and affective style." In Davidson and Hugdahl, pp. 361–388.

24 Bradshaw, John, and Rogers, Lesley. (1993). *The Evolution of Lateral Asymmetries, Language, Tool Use, and Intellect*. San Diego Academic Press.
p. 176. Ornstein, (1997). p. 75.

25 Peters, Michael. (1996). "Handedness and its relation to other indices of lateralization." In Davidson and Hugdahl. pp.183–214. Peters interviews with author (2001); comment to text (2004).

26 Peters (1996), p. 198.

27 Ornstein. p. 105.

28 Crone, John. (1999). "Left brain, right brain." *The New Scientist,* 2193.

29 Strauss, E., and Goldsmith, S. M. (1987). "Lateral preferences and cerebral speech dominance." *Cortex, 19,* 165–177. Peters, Michael, interviews with author. (2002, 2004). Peters (1996) in Davidson and

Hugdahl, pp. 183–214.

30 Kimura, D. (1993). *Neuromotor Mechanisms in Human Communication*. New York: Oxford University Press.

31 Peters, M., and Pang, J. (1992). "Do 'right-armed' lefthanders have different lateralization of motor control for the proximal and distal musculature?" *Cortex, 28*, 391–399.

32 Govind, C. K. (1992). "Claw asymmetry in lobsters: Case study in developmental neuroethology." *Journal of Neurobiology, 23*(10): 1423–1445. Govind correspondence with author. (2002). Input from Michael Peters (2004).

33 Peters and Harris, interviews with author (2004).

34 Peters, ibid.

35 Peters, ibid. See also Calvin, W. H. (1983). *The Throwing Madonna: Essays on the Brain*. New York: McGraw-Hill.

36 Peters, M. (2000). "Contributions of imaging techniques to our understanding of handedness." In Manas K. Mandal and G. Tiwari (Eds.), *Side-Bias: A Neuropsychological Perspective*. Dordrecht: Kluwer Academic Publishers, pp. 191–222.

37 Cabeza, R, and Nyberg, L. (2000). "Imaging Cognition II: An Empirical Review of 275 PET and fMRI Studies." *Journal of Cognitive Neuroscience, 12*, 1–47.

38 Hirsch, J. et al. (2000). "An integrated functional magnetic resonance imaging procedure for preoperative mapping of cortical areas associated with tactile, motor, language, and visual functions." *Neurosurgery, 47*(3): 711–722. Binder, J. et al. (1996). "Determination of language dominance using functional MRI: A comparison with the Wada test." *Neurology, 46*, 978-984.

39 Hirsch, Joy, Interviews with author (2001). Cabeza and Nyberg, pp. 1–47.

40 Hasson et al. (2001). "Vase or face? A neural correlate of shape-selective grouping processes in the human brain." *Journal of Cognitive Neuroscience, 13*, 744–753.

41 Christman, S. (1995). "Independence versus integration of right and left hemisphere processing: Effects of handedness." In F. Kitterle (Ed.), *Hemispheric Communication: Mechanisms and Models*. Hillsdale, NJ: Lawrence Erlbaum Associates. Christman, S. D. (2001). "Individual

differences in Stroop and local-global processing: A possible role of interhemispheric interaction." *Brain and Cognition, 45,* 97-118.

42 Hall, Stephen S. "Journey to the Center of My Mind." *The New York Times Magazine.* June 6, 1999.

43 Peters (2000). Binder et al. (1996).

44 Geschwind, Dan, interviews with author (1999, 2001, 2004).

45 McManus, Chris. (2002). *Right Hand, Left Hand. The Origins of Asymmetries in Brains, Bodies, Atoms and Cultures.* London: Weidenfeld and Nicolson, pp. 228–232.

46 Springer and Deutsch, pp. 44–45. Blakeslee, Sandra. "Workings of split brain challenge notions of how language evolved." *The New York Times,* November 26, 1996.

47 Hirsch, interviews with author (2002).

48 Porac, Clare et al. (1998). "Illness and accidental injury in young and older adult left- and right-handers: Implications for genetic theories of hand preference." *Developmental Neuropsychology, 14*(1): 157–172.

49 Siebner, H. R. et al. (2002). "Long-term consequences of switching handedness: A positron emission tomography study on handwriting in 'converted' left-handers." *The Journal of Neuroscience, 22*(7): 2816–2825.

50 Ibid.

51 Geschwind, Dan, interviews (2001, 2004). Peters interviews (2004). McManus (2002) ibid.

52 Geschwind, D.H. et al. (2002). "Heritability of lobar brain volumes in twins supports genetic models of cerebral laterality and handedness." *Proceeding of the National Academy of Sciences USA, 99,* 3176–3181.

53 Geschwind study, ibid. Annett, A., and Alexander, M. (1996). "Atypical cerebral dominance: Predictions and tests of the right-shift theory." *Neuropsychologia, 34*(12) 1215–1227. Dellatolas, et al. (1993). "Upper limb injuries and handedness plasticity." *British Journal of Psychology, 84*(May): pp. 201–205. (1993).

54 Horowitz, Janice M. "High five for a new hand; years after losing his hand in a fireworks accident, a patient is whole again—thanks to a daring transplant." *Time,* August 28, 2000.

55 Jäncke, L., Peters, M., et al. (1998). "Differential magnetic resonance signal change in human sensorimotor cortex to finger movements of

different rate of the dominant and subdominant hand." *Cognitive Brain Research*, 6(4): 279–284.

56 Amunts, K. et al. (1997). "Motor cortex and hand motor skills: Structural compliance in the human brain." *Human Brain Mapping, 5*, 206–215. Amunts, K. et al. (1996). "Asymmetry in the human motor cortex and handedness." *Neuroimage, 4*, 216–222.

57 Elbert, T. et al. (1995). "Increased cortical representation of the fingers of the left hand in string players." *Science, 270*, 305–307.

58 Witelson, S. F. (1985). "The brain connection: The corpus callosum is larger in left-handers." *Science, 229* (4714): 665–668. Beaton, A. A. "The relation of planum temporale asymmetry and morphology of the corpus callosum to handedness, gender, and dyslexia." (1997). *Brain and Language, 60*(2): 255–322. Clarke and Zaidel. (1994). "Anatomical-behaviorial relationships: corpus callosum morphometry and hemispheric speculation." *Behavioral Brain Research, 64*, 185–202. Cowell, Kertesz, and Denenberg. (1993). "Multiple dimensions of handedness and the human corpus callosum. *Neurology, 43*; 2353–2357.

59 Christman, S. D., and Propper, R. E. (2001). "Superior episodic memory is associated with interhemispheric processing." *Neuropsychology, 15*, 607–616. Nyberg, L. et al. (2000). "Large scale neurocognitive networks underlying episodic memory." *Journal of Cognitive Neuroscience, 12*, 163–173.

60 Christman, S. D. (1993). "Handedness in musicians: Bimanual constraints on performance." *Brain and Cognition, 22*, 266–272.

61 McCartney story: "The Listener: Strike a Chord; This Week's Selection of the Best of BBC Radio." London *Independent*, October 24, 1999. Hendrix story: *Rolling Stone*, August 26, 2003.

62 Christman. (2001). *Brain and Cognition*.

Chapter 5

1 Blau, Abraham. (1946). *The master hand; a study of the origin and meaning of right- and left-sidedness and its relation to personality and language*. Series title: American Orthopsychiatric Association. Research monographs. No. 5. p. 96. Harris, L. J. (2000). "On the evolution of handedness: A speculative analysis of Darwin's views and a review of

early studies of handedness in 'the nearest allies of man.'" *Brain and Language, 73,* 155.

2 Lawson, Iain. "Memories of Anthony Kerr, Gart Cosh Marcher." *The Scots Independent*, April 1987.

3 Anonymous. (1974). "The handedness of Kerrs—a surname study." *Journal of the Royal College of General Practitioners,* 24(143): 437–439.

4 Sicotte, N. L., Woods, R. P., and Mazziotta, J. P. (1999). "Handedness in twins: A meta-analysis." *Laterality, 4,* 265–286. Annett, M. (1978). "Genetic and nongenetic influences on handedness." *Behaviour Genetics, 8,* 227–249. Carter-Saltzman, L. (1980). "Biological and socio-cultural effects of handedness. Comparison between biological and adoptive families." *Science, 209,* 1263–1265.

5 Bello, Mark. (July 1986). "What causes left-handedness?" *Science, 7,* 83–85. Pye-Smith, Philip. H. (1871). "On left-handedness." *Guy's Hospital Reports, 16,* 141–146.

6 Coren, S., and Porac, C. (1977). "Fifty centuries of right-handedness: the historic record." *Science, 198,* 631–632. Faurie, C., and Raymond, M. (2004). "Handedness frequency over more than 10,000 years." Proceedings of the Royal Society of London, B271, S43–S45.

7 Peters, M. (1997). "Left and right in classical Greece and Italy." *Laterality, 2,* 3–6.

8 De Kay, J. T. (1994). *The Left-hander's Handbook.* New York: M. Evans and Co. Holder, M. K. (2002). "Gauche! Left-handers in society." (www.indiana.edu/~primate/lspeak.htm). Fincher, Jack. (1977). *Sinister People: The Looking-Glass World of the Left-Hander.* New York: G P Putnam's Sons, pp. 30–39. Corbis photos.

9 Coniff, Richard. "So, You Want to Be the Boss?" *Discover, 5,*(21): 72. May 1, 2000.

10 Bodmer, Walter, and McKie, Robin. (1994). *The Book of Man: The Human Genome Project and the Quest to Discover Our Genetic Heritage.* New York: Oxford University Press.

11 Kang, Y., and Harris, L. J. (2000). "Handedness and footedness in Korean college students." *Brain and Cognition,* 43(1–3): 268–274.

12 Peters, Michael. Correspondence with author (2001, 2004). Komai, T., and Fukuoka, G. (1934). "A study of the frequency of left-handedness and left-footedness in Japanese school children." *Human Biology, 6,* 33–42.

13 Harris, *Brain and Language*, p. 147.

14 Neimark, J., Cochran, T., and Dossey, L. "Nature's clones; research on twins." *Psychology Today*, July 1, 1997.

15 Previc, F. H. (1996). "Nonright-handedness, central nervous system and related pathology, and its lateralization: A reformulation and synthesis." *Developmental Neuropsychology, 12*(4): 443–515. Coren, *The Left-Hander Syndrome.*

16 Bakan, P., Dibb, G., and Reed, P. (1973). "Handedness and birth stress." *Neuropsychologia, 11*(3): 363–366. Bello, ibid. Geschwind, Dan, interviews with author (2001, 2004).

17 Geschwind, N., and Galaburda, A. M. (1985c). "Cerebral lateralization. Biological mechanisms, associations, and pathology: III. A hypothesis and a program for research." *Archives of Neurology, 42*(7): 634–654.

18 Ibid.

19 Ibid.

20 Okie, Susan. "Bush's thyroid condition diagnosed as Graves' disease." *The Washington Post*, May 10, 1991.

21 Grimshaw, G.M., Bryden, M. P., and Finegan, J. K. (1995). "Relations between prenatal testosterone and cerebral lateralization in children." *Neuropsychology, 9,* 68–70. Nicholls, M. E. R. and Forbes, S. (1996). "Handedness and its association with gender-related psychological and physiological characteristics." *Journal of Clinical and Experimental Neuropsychology, 18,* 905–910.

22 Annett, Marian, correspondence with author (2001). Levy, Jerre, interview with author (2000).

23 Bryden, M. P. (1993). "Perhaps not so sinister." *Contemporary Psychology, 38,* 71–72. Peters, M. (1996). "Hand preference and performance in lefthanders." In D. Elliott and E. A. Roy (Eds.), *Manual Asymmetries* (pp. 99–120). Boca Raton, FL: CRC Press. Harris, Lauren, interviews with author (2004).

24 Harris, L. J., and Peters, M, interviews with author (2001, 2004).

25 Ibid.

26 Bryden, M. P., McManus, I. C., and Bulman-Fleming, M. B. (1994). "Evaluating the empirical support for the Geschwind-Behan-Galaburda model of cerebral lateralization." *Brain and Cognition, 26*(2): 103–167. Dellatolas, G. et al. (1990). "An epidemiological reconsideration of the

Geschwind-Galaburda theory of cerebral lateralization." *Archives of Neurology,* 47(7): 778–782. Porac, C., and Friesen, I. C. (2000). "Hand preference side and its relation to hand preference switch history among old and oldest-old adults." *Developmental Neuropsychology,* 17, 225–239.

27 Geschwind, D., interview with author (2004).

28 Ward, Jeannette. Interview with author (2001).

29 Flannery, K. A., and Liederman. J. (1995). "Is there really a syndrome involving the co-occurrence of neurodevelopmental disorder, talent, non-right handedness and immune disorder among children?" *Cortex, 31*(3): 503–515. Bryden et al. (1994). Anonymous. (April 1992). "The left-handed riddle." *Health News.* University of Toronto.

30 Porac, C., correspondence with author (2004).

31 Steenhuis, R. E. et al. (1990). "Reliability of hand preference items and factors." *Journal of Clinical and Experimental Neuropsychology, 12,* 921–930. Bryden et al. (1994). Ward, J. Interviews with author (2001). Biederman J. et al. (1995). "No confirmation of Geschwind's hypothesis of associations between reading disability, immune disorders, and motor preference in ADHD; attention deficit hyperactivity disorder." *Journal of Abnormal Psychology, 23*(5): 545.

32 Peters, M., and Ivanoff, J. (1999). "Performance asymmetries in computer mouse control for righthanders, and lefthanders with left- and right-handed mouse experience." *Journal of Motor Behavior, 31,* 86–94.

33 Marks, J. S., and Williamson, D. F. (1991). "Left-handedness and life expectancy." *New England Journal of Medicine, 325,* 1042. Rothman, K. J. (1991). "Left-handedness and life expectancy." *New England Journal of Medicine, 325,* 1041. Murphy, James J. "Long live lefties." *The New York Times,* February 15, 1992. Salive, M. E., Guralnik, J. M., and Glynn, R. J. (1993). "Left-handedness and mortality." *American Journal of Public Health." 83,* 265–267. Harris, Lauren J. (1993). "Do left-handers die sooner than right-handers? Commentary on Coren and Halperin's (1991). 'Left-Handedness: A Marker for Decreased Survival Fitness.' *Psychological Bulletin, 114*(2): 203–234. Annett, M. (1993). "The fallacy of the argument for reduced longevity in left-handers." *Perceptual and Motor Skills, 76,* 295–298. Murray, David. "Exit, stage left," *The National Post* (Canada), August 18, 2000.

34 Coren, S. (1992). *The Left-Hander Syndrome.* New York: The Free

Press, p. 210. Coren, S. and Halpern, D. F. (1991). "Left-handedness: a marker for decreased survival fitness." *Psychological Bulletin, 109*(1): 90–106. Halpern, D. F., and Coren S. (1988). "Do right-handers live longer?" *Nature, 3*(33): 213.

35 Salive et al. Marks et al.

36 Porac, C., & Friesen, I. C. (2000). "Hand preference side and its relation to hand preference switch history among old and oldest-old adults." *Developmental Neuropsychology, 17*, 225–239.

37 Annett, M. (1978). "Genetic and nongenetic influences on handedness." *Behaviour Genetics, 8*, 227–249. Annett, M. (1979). "Familial handedness in three generations predicted by the right shift theory." *Annals Human Genetics, 42*, 479–491.

38 Peters, M., and Perry, R. (1991). "No link between left-handedness and maternal age and no elevated accident rate in left-handers." *Neuropsychologia, 29*, 1257–1259. Smith, Barry et al. Hicks et al. (1993). "Handedness and accidents with injury." *Perceptual and Motor Skills, 77*(3): 1119–1122. Correspondence with Hicks (2000). Murphy, James J. "Long live lefties." *The Washington Post.* February 15, 1992. Shute, Nancy. "Southpaw Reprieve," *United States News & World Report*, October 9, 2000. Porac, C et al. (1998). "Illness and accidental injury in young and older adult left- and right-handers: Implications for genetic theories of hand preference." *Developmental Neuropsychology, 14*(1): 157–172.

39 Harris. (2000). *Brain and Language*, p. 155.

40 Hepper, P. G., Shahidullah, S., and White, R. (1991). "Handedness in the human fetus." *Neuropsychologia, 29*, 1107–1111.

41 Geschwind, D. H. et al. (2002). "Heritability of lobar brain volumes in twins supports genetic models of cerebral laterality and handedness." *Proceedings of the National Academy of Sciences, 99*, 3176–3181. Hotz, Robert Lee. "Left-handers are found to have different brains." *Los Angeles Times*, March 5, 2002.

42 Geschwind, D., ibid.

43 Annett, M. (1972). "The distribution of manual asymmetry." *British Journal of Psychology, 63*(3): 343–358.

44 Annett (1972), ibid. Annett correspondence with author. (2001). (See also: Annett, M. (2002). *Handedness and Brain Asymmetry: The Right*

Shift Theory. Hove, UK: Psychology Press.)

45 Annett (1972), ibid. Peters, M, correspondence with author (2004).

46 Geschwind, D. (2002), ibid. Author interviews with Geschwind (2004).

47 Geschwind, D. (2002), ibid. Author interviews with Peters.

48 Francks C. et al. (2003). "Confirmatory evidence for linkage of relative hand skill to 2p12-q11." *American Journal of Human Genetics, 72,* 499–502. Author correspondence with Francks (2004).

49 Levy, Jerre, interview with author (2001). (See also: Levy, J. (1976). "A review of evidence for a genetic component in the determination of handedness." *Behavior Genetics, 6,* 429–453.) Geschwind, D., interviews with author (2001, 2004).

50 Geschwind, D. H., and Miller, B. L. (2001). "Molecular approaches to cerebral laterality: Development and neurodegeneration." *American Journal of Medical Genetics, 101,* 370–381

51 Geschwind, D. (2001), ibid.

52 Laland, K. N. et al. (1995). "A gene-culture model of human handedness." *Behavior Genetics, 25,* 433–445.

53 Author interviews with Michael Peters, Clare Porac, Stephen Christman, Daniel Geschwind, Marian Annett (2000–2004).

54 McManus, Chris. (2002). *Right Hand, Left Hand: The Origins of Asymmetry in Brains, Bodies, Atoms and Cultures*. London: Weidenfeld and Nicolson, pp. 207–209.

55 Ibid., p. 209.

56 McManus, I. C., and Bryden, M. P. (1992). "The genetics of handedness, cerebral dominance and lateralization." In I. Rapin and S. J. Segalowitz (Eds.), *Handbook of Neuropsychology, Volume 6: Developmental Neuropsychology* Amsterdam: Elsevier Science. (pp. 115–144).

Chapter 6

1 Dwight, Thomas. (1891). "What is right-handedness?" *Scribner's, 9,* 465–476. From Harris, L. J. (2000). "On the evolution of handedness: A speculative analysis of Darwin's views and a review of early studies of handedness in 'the nearest allies of man.'" *Brain and Language, 73,* 156.

2 Gibbons, Ann. (1991). "Deja vu all over again: Chimp-language wars."

Science, 251(5001): 1561. Ward, J. Interviews with author (2000, 2001, 2004).

3 Ornstein, Robert. (1991). *The Evolution of Consciousness: Of Darwin, Freud, and Cranial Fire—The Origins of the Way We Think*. New York: Prentice Hall, pp. 50–51.

4 Darwin, C. (1871). *The Descent of Man and Selection in Relation to Sex* (2 vols.). London: John Murray, pp. 430–444.

5 Calvin, William H. (1983). "A stone's throw and its launch window: Timing precision and its implications for language and hominid brains." *Journal of Theoretical Biology, 104,* pp. 121–135. Calvin, W. H. (1993). "The unitary hypothesis: A common neural circuitry for novel manipulations, language, plan-ahead, and throwing?" In K. Gibson and T. Ingoldm, *Tools, Language, and Cognition in Human Evolution*. Cambridge: Cambridge University Press, pp. 230–250. (See also: www. williamcalvin.com.)

6 Peters, Michael. Interviews/correspondence with author (2001, 2004). Calvin (1993).

7 Calvin (1993). Ornstein (1991), ibid.

8 Chiseling/steadying: Ambrose, Stanley. (2001). "Paleolithic technology and human evolution." *Science*, March 2, 2001. No. 5509, Vol. 291, p. 1748. Evidence of right hand dominance, 1.5 million years ago: Toth, Nicholas. "Making Silent Stones Speak."(1985). *Journal of Human Evolution, 14,* 607. Handedness in recent epochs: Faurie, C. and Raymond, M. (2004). "Handedness frequency over more than 10,000 years." Proceedings of the Royal Society of London B 271, S43–S45.

9 Corballis, Michael C. (1999). "The Gestural Origins of Language." *American Scientist, 87,* p. 138. (See also: Corballis, Michael. (1991). *The Lopsided Ape*. Oxford University Press.)

10 Corballis, ibid. Annett, M. (1985). *Left, Right, Hand and Brain: The Right Shift Theory*. Hillsdale, NJ: Lawrence Erlbaum and Associates. Annett correspondence with author (2001).

11 Corballis, correspondence with author (2001).

12 Corballis, ibid. Poizner, H. et al. (1987). *What the Hands Reveal About the Brain*. Cambridge: MIT Press.

13 Crow, T. J. (1998). "Sexual selection, timing and the descent of man: A theory of the genetic origins of language." *Cahiers de Psychologie*

Cognitive/ *Current Psychology of Cognition, 17,* 1079–1114. Crow, correspondence with author (2002). Darwin, *Descent.*

14 Faurie and Raymond (2004).

15 Hopkins, W. D. (1996). "Chimpanzee handedness: Causes and consequences." *International Journal of Psychology, 31, 365.* Ward, Jeannette. Interviews with author (2001).

16 Marchant, L. F., and McGrew, W. C. (1996). "Laterality of limb function in wild chimpanzees of Gombe National Park: comprehensive study of spontaneous activities," *Journal of Human Evolution, 30,* 427–443. MacNeilage, P. interviews with author (2001). Hopkins, W. D. (1998). "Maternal influences on the development of hand preference in infant chimpanzees." *Infant Behavior and Development, 21,* 111. Hopkins, Bill, interviews with author (2001, 2004).

17 Begley, Sharon. "Aping language." *Newsweek,* January 19, 1998. Morell, Virginia. (October 4, 1991). "A hand on the bird—and one on the bush: a controversial new theory holds that nonhuman primates are 'handed' just as humans are." *Science, 254*(5028): 33–36. Gannon, Patrick J. (January 9, 1998). "Asymmetry of chimpanzee planum temporale: Humanlike pattern of Wernicke's brain language area homolog." *Science,* (5348) 279: 220.

18 Gannon, ibid. Begley, ibid.

19 Begley, ibid. MacNeilage, interviews with author (2001).

20 Calvin (1993). Morell (1991).

21 Bradshaw, John, and Rogers, Lesley. (1993). *The Evolution of Lateral Asymmetries, Language, Tool Use, and Intellect,* p. 176.

22 Sanford, C., and Ward, J. P. (1986). "Mirror image discrimination and hand preference in the bush baby (*Galago senegalensis*)." *Psychological Record, 36,* 439–449.

23 Ward, J. P. (1988). "Left hand reaching preferences in prosimians." *Behavioral and Brain Sciences, 11,* 732–733. Ward, J. P., and Cantalupo, C. (1997). "Origins and functions of laterality: Interactions of motoric systems." *Laterality, 2,* 279–303. Ward, J. P. (1999). "Left hand advantage for prey capture in the galago (Galago moholi)." *International Journal of Comparative Psychology, 11*(4): 173–184.

24 MacNeilage, P. F., Studdert-Kennedy, M. G., and Lindstrom, B. (1987). "Primate handedness reconsidered." *Behavioral and Brain Sciences, 10;*

247–303. MacNeilage, interviews with author (2001, 2004).

25 Hopkins, W. D. (1993). "Explaining variability in chimpanzee (Pan troglodytes) hand preference: Task, posture, genetic and ontogenetic factors." *American Journal of Primatology, 30*, 317. Hopkins, interviews with author (2001, 2004).

26 MacNeilage, P. F. (1991). "The postural origins' theory of primate neurobiological asymmetries." In N. A. Krasnegor, et al. *Biological and Behavioral Determinants of Language Development.* Mahwah, NJ: Lawrence, Erlbaum and Associates. pp. 165–187. MacNeilage, interviews with author (2000–2001).

27 MacNeilage, ibid.

28 Ward interviews/correspondence with author (2001, 2004).

29 Ward ibid. Tosches, Rich. "Lefties: Make no mistake, it's a right-hander's world. Except in sports—where being out of step can put a player in sync." *Los Angeles Times*, December 4, 1986.

30 MacNeilage, interviews with author (2001, 2004).

31 Davidson, R. J., et al. (1990). "Approach-withdrawal and cerebral asymmetry: emotional expression and brain physiology." *Journal of Personality and Social Psychology, 58*, 330–341.

32 Bradshaw and Rogers, pp. 25–26. Govind, C. K., correspondence with author (2001). Bisazza et al. (1996). "Right-pawedness in toads," *Nature, 379.* Naitoh, T., and Wasserung, R. (1996). "Why are toads right-handed?" *Nature, 380.*

33 Rogers, Lesley J. (2000). "Evolution of hemispheric specialization: Advantages and disadvantages." *Brain and Language, 73*, 236–253.

34 Bradshaw and Rogers, pp. 38–42.

35 Bradshaw and Rogers, pp. 42–50. Vallortigara, G. et al. (1998). "Complementary right and left hemisfield use for predatory and agonistic behaviour in toads." *NeuroReport, 9*, 3341–3344.

36 Bradshaw and Rogers, ibid. Vallortigara, ibid.

37 Bradshaw and Rogers, p. 51.

38 Vallortigara, G. (2000). "Comparative neuropsychology of the dual brain: A stroll through animals' left and right perceptual worlds." *Brain and Language, 73*, 189–219. Hoptman, J. M., and Levy, J. (1988). "Perceptual asymmetries in left- and right-handers for cartoon and real faces," *Brain and Cognition, 8:*

178–188. Levy, Jerre, interviews with author (2000).

39 Vallortigara, G. et al. (1999). "Detour behaviour, imprinting and visual lateralization in the domestic chick." *Cognitive Brain Research, 7*, 307–320.

40 Annett, M. (1985). Annett, M., and Manning, M. (1990). "Reading and a balanced polymorphism for laterality and ability." *Journal of Child Psychology and Psychiatry 31*(4): 511.

41 Annett, (1990). ibid.

42 McManus, I. C., Shergill, S., Bryden, M. P. (1993). "Annett's theory that individuals heterozygous for the right shift gene are intellectually advantaged: Theoretical and empirical problems; comment on an article by Marian Annett in this issue." (p. 511). *British Journal of Psychology, 84*(4): 517.

Chapter 7

1 Nicholl, Charles. (2004). *Leonardo da Vinci: Flights of the Mind*. New York: Viking, p.325. Aaron, P. G., and Clouse, R. G. (1982). "Freud's psychohistory of Leonardo de Vinci: A matter of being right or left." *The Journal of Interdisciplinary History, 13*(1): 1–16.

2 Calvin, W. H. (1993). "The unitary hypothesis: A common neural circuitry for novel manipulations, language, plan-ahead, and throwing?" In K. Gibson and T. Ingoldm, *Tools, Language, and Cognition in Human Evolution*. Cambridge: Cambridge University Press, pp. 230–250. (See also: WilliamCalvin.com.) Peters, Michael, correspondence with author (2004).

3 Fraley, Gerry. "Lefthanders just aren't normal folks—and here's why." *Dallas Morning News*, April 2, 2000.

4 Elliott, Digby. (1993). "Asymmetries in the preparation and control of manual aiming movements." *Canadian Journal of Experimental Psychology, 47*(3): 570–589. Elliott, interview with author (2001).

5 Escamilla, Raphael, interview with author (2001).

6 Annett, Marian, correspondence with author (2001). (See also: Annett, M. [2002]. *Handedness and Brain Asymmetry: The Right Shift Theory*. Hove, U. K.: Psychology Press.

7 Coren, S. and Porac, C. (1982). *American Journal of Optometry and*

Physiological Optics, 59, 987–990. Elliott. (1993).

8 Dane, S., and Erzurumluoglu, A. (2003). "Sex and handedness differences in eye-hand visual reaction times in handball players." *International Journal of Neuroscience, 113;* 923–929.

9 Lee, W. A. (1980). "Anticipatory control of postural and task muscles during rapid arm flexion." *Journal of Motor Behaviour, 12,* 185–196.

10 The National Sports Gallery.

11 Newnham, David. "When words dance." *The Guardian* (London), April 24, 1999.

12 Direction of queen on pound: Martin, M., and Jones, G. V. (1997). "Handedness dependency in recall from everyday memory." *British Journal of Psychology,* 4(88): 609–619. Nomadic people: Dawson, John L. (1977). "An anthropological perspective on the evolution and lateralization of the brain." *Annals of the New York Academy of Science, 299,* 440.

13 Bradshaw, John and Rogers, Lesley. (1993). *The Evolution of Lateral Asymmetries, Language, Tool Use, and Intellect,* San Diego: Academic Press.
pp. 38–42.

14 Heilman, K. M. et al. (1987). "Attention: Behavior and neural mechanisms." In V. B. Mountcastle et al (Eds.), *Handbook of Physiology. Higher functions of the brain.* American Physiological Society (V), pp. 461–481. Hoptman, J. M., and Levy, J. (1988). "Perceptual asymmetries in left- and right-handers for cartoon and real faces." *Brain and Cognition, 8,* 178–188.

15 Ornstein, Robert. (1997). *The Right Mind. Making Sense of the Hemispheres.* New York: Harcourt Brace, p. 27.

16 Ekman, P., and O'Sullivan, M. (1991). "Who can catch a liar?" *American Psychologist,* 46(September): 913–920. Porter, Stephen et al. (2002). "The influence of judge, target, and stimulus characteristics on the accuracy of detecting deceit." *Canadian Journal of Behavioural Science,* 34(3): 172–180.

17 McManus, Chris. (2002). *Right Hand, Left Hand: The Origins of Asymmetries in Brains, Bodies, Atoms and Cultures.* London: Weidenfeld and Nicolson, p. 231. Psychiatrists: Ransil, B. J., and Schachter, S. C. (1996). "Inventory-derived task handedness preferences of nine

professions and their associations with self-report global handedness preferences." *Perceptual and Motor Skills, 86,* 303–320. Aaron and Clouse. (1982). pp. 1-16.

18 Corballis, Michael C. "The Gestural Origins of Language." *American Scientist,* March 1, 1999, 138. (See also: Corballis, Michael. [1991] *The Lopsided Ape.* Oxford University Press.) Ransil, ibid. Fincher, Jack. (1977) *Sinister People: The Looking-Glass World of the Left-Hander,* New York: G P Putnam's Sons, pp. 30–39. Holder, M. K. (2002). "Gauche! Left-handers in society." (www.indiana.edu/~primate/lspeak. html).

19 Ornstein (1997). *The Right Mind,* p. 76. Drake, Roger. (2003). "Neurobiology of Optimism." Presentation to the Gallup Institute, Hungary. Sontam, V. et al. (2004). "Neuropsychological determinants of risk perception." Presented at the 45th Annual Meeting of the Psychonomic Society, Minneapolis.

20 Sontam et al. (2004).

21 Davidson, R. J., and Rickman, M. (1999). "Behavioral inhibition and the emotional circuitry of the brain: Stability and plasticity during the early childhood years." In R. J. Davidson et al. *Series in Affective Science. Extreme Fear, Shyness, and Social Phobia: Origins, Biological Mechanisms, and Clinical Outcomes.* Oxford University Press, pp. 67–87.

22 Ornstein, R. (1997). pp. 97–103. Coulson, S., and Lovett, C. (2004) "Handedness, hemispheric asymmetries, and joke comprehension." *Cognitive Brain Research, 19*(3): 275–288.

23 Merekelbach et al. (1989). "Handedness and anxiety in normal and clinical populations." *Cortex, 25*(4): 599–606. Giotakos, O. (2001). "Narrow and broad definition of mixed-handedness in male psychiatric patients." *Perceptual and Motor Skills, 93,* 631–638. Author correspondence with Christman (2005).

24 Bello, Mark. (July 1986). "What causes left-handedness?" *Science, 7,* 83–85. Dawson. (1977). Ibid.

25 Cashman anecdote: Rutledge, Leigh. (2002). *The Left-Hander's Book of Days.* New York: Dutton.

26 Hori, M. (1993). "Frequency-dependent natural selection in the handedness of scale-eating Cichlid fish." *Science, 260,* 216–219.

Raymond, M. et al. (1996). "Frequency dependent maintenance of left handedness in humans." *Proceedings of the Royal Society of London B, 263,* 1627–1633. Faurie, C., Pontier, D., and Raymond, M. (2004). "Student athletes claim to have more sexual partners than other students." *Evolution and Human Behavior, 25,* 1–8.

27 Grouios, G. et al. (2000). "Do left-handed competitors have an innate superiority in sports?" *Perceptual and Motor Skills, 90*(3, PT2): 1273–1282.

28 Holtzen, D. W. (2000). "Handedness and professional tennis." Gordon and Breach Science. Author interview with Holtzen (2000).

29 Azemar, G. (1993). "Les gauchers en escrime: Donnés, statistique et interpretation." *Escrime Internationale, 7,* 15–19.

30 Holtzen, interview with author (2000).

31 Hotzen, ibid. Sanders, B., Wilson, J. R., and Vandenberg, S. G. (1982). "Handedness and spatial ability." *Cortex, 18,* 79–90.

32 Azemar. (1993).

33 Kilshaw, D., and Annett, M. "Right and left-hand skill. I.: Effects of age, sex, and hand preference showing superior skill in left-handers." *British Journal of Psychology, 74,* 253–268. Author correspondence with Annett (2000–2001). Dane and Erzurumluoglu (2003).

34 Gatward, Matt. "Sport: The charge of the left brigade." *The Independent* (London), July 27, 1999, p. 23. Azemar, G. et al. (1983). "Etude neuropsychologique du comportement des gauschers en escrime." *Cinesiologie, 22,* 7–18.

35 Holtzen, interview with author (2000).

36 Holder, M. K. (2002). Fincher (1977). Rutledge and Donley (1995).

37 Roth, Melissa. "Left wing politics." *The New York Times,* January 20, 2000. Author correspondence with Galaburda (2000).

38 Shute, Nancy. (December 1994). "Life for lefties: From annoying to downright risky." *The Smithsonian, 25*(9) 130–143.

39 Cranberg, L., and Albert M. In L. Obler and D. Fein (Eds.), *The Exceptional Brain.* New York: Guilford.

40 Roth (2000). Author interview with Coren (2000).

41 Trotter, Robert J. "Sinister psychology." *Science News,* October 5, 1974, p. 221.

42 Roth (2000). Author correspondence with Gardner (2000).

43 Corbis photo archives.

44 ABC News, *Primetime Live:* "Left-handed candidates." October 1, 1992. Coren, S. (1995). "Differences in divergent thinking as a function of handedness and sex." *American Journal of Psychology, 108*(3): 311–325.

45 Sanderson, Rachel. "Italy show cracks the code of da Vinci's mind." MSNBC.com/Reuters. January 5, 2005.

46 Barnett, K. J. and Corballis, M. C. (2002). "Ambidexterity and magical ideation." *Laterality, 7,* 75–84. Witelson, S. F., Kigar, D. L., and Harvey, T. (1999). "The exceptional brain of Albert Einstein." *The Lancet, 353,* 2149.

47 Nicholl, Charles. (2004). p. 417.

48 Ibid.

49 White, Michael. (1999). *Leonardo: The First Scientist.* New York: Warner Books.

50 Aaron and Crouse. (1982).

51 Christman, correspondence with author (2005).

52 McManus (2002), p. 208.

53 Ibid.

54 Faurie, C., Pontier, D., and Raymond, M. (2004). "Student athletes claim to have more sexual partners than other students." *Evolution and Human Behavior, 25;* 1–8.

55 Dawson, J. L. M. (1974). "Ecology, cultural pressures towards conformity and left-handedness: A bio-social approach." In J. L. M. Dawson and W. J. Lonner (Eds.), *Readings in Cross-cultural Psychology.* Aberdeen: Hong Kong University Press, pp. 130–136.

56 Miller, Geoffrey F. (2001). "Aesthetic fitness: How sexual selection shaped artistic virtuosity as a fitness indicator and aesthetic preferences as mate choice criteria." *Bulletin of Psychology and the Arts, 2*(1); 20–25. (See also: Miller, Geoffrey F. [2001]. The mating mind: How sexual choice shaped the evolution of human nature, *Psycology, 12*[8].)

57 Peterson, J. M., and Lansky, L. M. (1977). "Left-handedness among architects: Partial replication and some new data." *Perceptual and Motor Skills, 45*(3 Pt 2): 1216–1218. Schacter, S. C. and Ransil, B. J. (1996). "Handedness distributions in 9 professional groups." *Perceptual and Motor Skills, 82,* 51–63.

58 List of musicians: Holder, M. K. Indiana University. Corbis photo
research. Author correspondence with Christman (2004).

59 Christman, S. D. (1993). "Handedness in musicians: Bimanual
constraints on performance." *Brain and Cognition, 22,* 266–272.

60 Miller, G. F. (2000). "Evolution of human music through sexual
selection." In N. L. Wallin, B. Merker, and S. Brown (Eds.), *The Origins
of Music.* Cambridge, MA: MIT Press, pp. 329–360.

61 Miller, G. F. (2000) ibid. Small, M. (1993). *Female Choices: Sexual
Behavior of Female Primates.* Ithaca, NY: Cornell University Press.

62 Miller, G. F. (2000), ibid. Darwin, C. (1871). *The Descent of Man,
and Selection in Relation to Sex* (2 vols.). London: John Murray, pp.
590–595.

63 Miller, G. F. (1997). "Protean primates: The evolution of adaptive
unpredictability in competition and courtship." In A. Whiten and R. W.
Byrne (Eds.), *Machiavellian Intelligence II: Extensions and Evaluations*
(pp. 312–340). Cambridge, England: Cambridge University Press.

64 Levy, interview with author (2000).

65 Holtzen, interview with author (2000).

66 Christman, interview with author (2005).

Additional sources and information may be found at www.theleftstuff.com.

Additional Resources

Books

Annett, Marian. *Handedness and Brain Asymmetry: The Right Shift Theory.*

DeKay, James. *The Left-Hander's Handbook*

Healy, Jane. *Loving Lefties: How to Raise Your Left-Handed Child in a Right-Handed World.*

McManus, Chris. *Right Hand, Left Hand: The Origins of Asymmetries in Brains, Bodies, Atoms and Cultures.*

Paul, Diane. *Left-Handed Helpline: An Essential Guide for Teachers, Teacher Trainers and Parents of Left-Handed Children.*

Rutledge, Leigh. *The Left-Hander's Book of Days.*

Rutledge, Leigh, and Richard Donley. *The Left-Hander's Guide to Life: A Witty and Informative Tour of the World According to Southpaws.*

Websites

M.K. Holder, PhD.

http://handedness.org/action/leftwrite.html

Includes tips for teaching left-handers to write.

Anything Left-Handed.com

www.Left-handersDay.com

Left-handed.com

Rosemary West's Left-Handed World

www.rkwest.com/left/index.shtml

Index

Hertz, Robert, 26, 28
Hicks, Robert A., 54
Hinduism, 27, 29-30
Hippocrates, 64-65
Hirsch, Joy, 63, 65, 89
Holtzen, David W., 49, 176, 195
Hopkins, Bill, 131, 140
Human balanced polymorphism theory,
 155
Human Performance Laboratory, 163
Humor, 173
Hygiene, 32

I
Insanity, 34
Intellectually challenged, 114
Islam, 33

J
Jamblichus, 27
James I (of England), 30
Judeo-Christian traditions, 31
K
Kabbalah, 31
Kang, Yeonwook, 57
Kuhl, Patricia, 140

L
Laland, Kevin, 53, 127
Lane, E.W., 33
Language Learning Center (Atlanta),
 143
Laterality quotient, 49
Lateralization, 19, 51, 59
Lee, Win, 167
Left-hander syndrome, 13, 17, 35, 45
Leigh, Adam, 19
Levy, Jerre, 111, 124, 153, 194
Life spans, 17–19, 108, 115
Lombroso, Cesare, 33

M
MacNeilage, Peter, 141, 144, 148, 166
Magical ideation, 35
Maori, 25, 28
Marchant, Linda, 141
McGrew William, 141
McKie, Robin, 105
McManus, Chris, 25, 92, 112, 128, 188
McMaster University (Canada), 164
Miami University (Ohio), 141
Michigan State University, 35, 57, 107
Migraine headaches, 12
Military leadership, 175, 179–180
Miller, Geoffrey, 190
Mixed-handers, 35, 48, 51–54, 59, 185.
 see also Switching
Musicians, 99, 191
Muslims, 33, 56

N
National Cancer Institute (Maryland),
 50
National Health and Nutrition
 Examination Survey, 116
National Institute of Sports and
 Physical Education (Paris), 177
National Institute on Aging, 19, 115,
 116

O
Ornstein, Robert, 80, 133, 170
Oxford University, 123, 139, 168

P
Palmer, A.N., 36
Parmenides, 32
Pathological left-handedness theory,
 107
Pennsylvania State University, 113
Performance, 45, 47, 48, 51, 54
Peters, Michael, 53, 78, 84-85, 96, 161
Piazza, Mike, 166

About the Author

Melissa Roth is a left-handed, right-footed freelance journalist who has written for publications including *The New York Times, Rolling Stone,* and *Marie Claire.* Though her early lefty predilections led teachers to send her for dyslexia testing, she managed to graduate from Cornell and Columbia Universities. She currently lives in New York City.